Introduction to OCEANOGRAPHY

Introduction to OCEANOGRAPHY

Charlie Jones
Emily O'Brien
Alexander Taylor

Kruger Brentt
Publishers
2023

Kruger Brentt Publishers UK. LTD.
Company Number 9728962

Regd. Office: 68 St Margarets Road, Edgware, Middlesex HA8 9UU

ISBN: 978-1-78715-093-5

For information on all our publications visit our website at http://krugerbrentt.com/

PREFACE

Oceanography, scientific discipline concerned with all aspects of the world's oceans and seas, including their physical and chemical properties, their origin and geologic framework, and the life forms that inhabit the marine environment. Traditionally, oceanography has been divided into four separate but related branches: physical oceanography, chemical oceanography, marine geology, and marine ecology. Physical oceanography deals with the properties of seawater (temperature, density, pressure, and so on), its movement (waves, currents, and tides), and the interactions between the ocean waters and the atmosphere. Chemical oceanography has to do with the composition of seawater and the biogeochemical cycles that affect it. Marine geology focuses on the structure, features, and evolution of the ocean basins. Marine ecology, also called biological oceanography, involves the study of the plants and animals of the sea, including life cycles and food production.

Oceanography is the sum of these several branches. Oceanographic research entails the sampling of seawater and marine life for close study, the remote sensing of oceanic processes with aircraft and Earth-orbiting satellites, and the exploration of the seafloor by means of deep-sea drilling and seismic profiling of the terrestrial crust below the ocean bottom. Greater knowledge of the world's oceans enables scientists to more accurately predict, for example, long-term weather and climatic changes and also leads to more efficient exploitation of the Earth's resources. Oceanography also is vital to understanding the effect of pollutants on ocean waters and to the preservation of the quality of the oceans' waters in the face of increasing human demands made on them.

The present book contains eleven -chapters covering all related disciplines. These chapters include 13 chapters. These chapters include Introduction, Introduction to Planet "Earth", Plate Tectonics and the Ocean Floor, Marine Sediments, Water and Seawater , Air-Sea Interaction, Ocean Circulation, Waves and Water Dynamics, Tides, Beaches, Shoreline Processes and the Coastal Ocean, Marine Pollution, Marine Life and the Marine Environment, Animals of the Pelagic Environment. Related terminology is given at the end for ready reference. It is written in a lucid style with the most recent advances, and with theoretical discussions being supplemented with illustrations, and tables for easy understanding of the subject. In this book most of the relevant topics are covered adequately. The book is also supported with a comprehensive glossary and index in the end. This book is written for students in meteorology, ocean engineering, and oceanography.

We are grateful to all those persons as well as various books, manuals, periodicals, magazines, journals etc. that helped in the preparation of this book. In spite of the best efforts, it is possible that some errors may have occurred into the compilation and editing of the book. Further queries, constructive suggestions and criticisms for the improvement of the book are always welcome and shall be thankfully acknowledged.

Charlie Jones
Emily O'Brien
Alexander Taylor

CONTENTS

CHAPTER

INTRODUCTION 1

The oceans are a major contributor to global climate patterns, and can give us clues to past climate conditions. Knowledge of oceanography is important for resource or energy extraction, such as commercial fishing or aquaculture, oil and gas exploration, and clean energy resources such as wind, wave, or tidal energy. The oceans are the major route for international trade through commercial shipping, and are still a significant factor in the transportation of people across the seas.

But even the recreational user benefits from an understanding of the ocean, from winds and currents for the casual sailor, tides and habitat conditions for a fisherman, to wave patterns for surfers. And finally, for anyone who has ever stood on the shore and gazed out at the ocean with a sense of wonder at what lies beneath the surface, the study of oceanography can begin to reveal some of the ocean's mysteries.

IMPORTANCE OF THE OCEAN TO SOCIETY

The ocean dominates Earth's surface and greatly affects our daily lives. It regulates Earth's climate, plays a critical role in the hydrological cycle, sustains a large portion of Earth's biodiversity, supplies food and mineral resources, constitutes an important medium of national defense, provides an inexpensive means of transportation, is the final destination of many waste products, is a major location of human recreation, and inspires our aesthetic nature.

Today's sense of urgency about ocean studies is precipitated by human impacts on oceanic systems and the need for a better understanding of the ocean's role in controlling global chemical, hydrological, and climate processes.

The nation is faced with pressing marine research problems whose timely solution will require increased cooperation between federal agencies and academic scientists. Many of these problems arise from the need to accommodate multiple uses of the ocean and from the ever-increasing concentration of the U.S. population near its coasts. Oceanographic research is important to many of the nation's social concerns, including the following:

Global Change

The ocean plays a predominant role in regulating both natural and human-induced changes in our planet. The role of ocean circulation and the coupling of the ocean and the atmosphere are basic to understanding Earth's changing climate. Regional events such as El Niño and ocean margin and equatorial upwelling influence climate on both seasonal and longer time scales. The world's population is now large enough to alter the chemical composition of the ocean and atmosphere and to impact the biological composition of Earth.

Biodiversity

The oceans comprise a large portion of Earth's biosphere and support a vast diversity of flora and fauna that are critical to Earth's biogeochemical cycles and serve as an important source of food and pharmaceuticals. In addition to the exciting discoveries of previously unknown biota near hydrothermal vents, many deep-ocean organisms have evolved under relatively stable conditions. Their unique physiologies and biochemistries have not yet been explored adequately, and methods for sampling the more fragile of these species have been developed only in the past decade.

Human influence on marine biota has increased dramatically, threatening the stability of coastal ecosystems. Some species have been overharvested; others have been transported inadvertently to areas where they are not indigenous, sometimes resulting in deleterious effects on native species.

Still other species are being cultivated commercially, and aquaculture facilities along coastlines are becoming commonplace in some countries. A better understanding of the ecology of marine organisms is urgently needed to prevent irreversible damage to this living resource.

Environmental Quality

Waste disposed in coastal areas has reached the open ocean, with broad ramifications for living resources. This problem is compounded because many marine species harvested for commercial and recreational purposes spend a portion of their lives in coastal waters and estuaries. Thus, local pollution can have far-reaching effects.

Economic Competitiveness

Economic prosperity in a global marketplace depends increasingly on technical and scientific applications. There is concern about the ability of the United States to compete with Europe and Asia. Basic and applied research in marine science and engineering is necessary to achieve and maintain a competitive position in a host of fields, including marine biotechnology, aquaculture, hydrocarbon and mineral exploration and production, maritime transportation, fisheries, treatment and disposal of waste, and freshwater extraction.

National Security

Unprecedented world political changes are redefining national defense interests and altering research and development priorities. Knowledge of the ocean, especially the acoustic properties of marginal seas and coastal areas, is critical to national defense. Experience gained in 1991 during the war in the Persian Gulf highlights the need for better information related to oceanic and coastal processes and to maritime operations and transportation.

Energy

The ocean's energy resources are essential to the national economy and national security. After a decade of relative neglect, energy issues are reemerging. With oil supplies continually threatened by instability in the Middle East and with increasing atmospheric carbon dioxide viewed as a possible trigger of global warming, there is a need to look carefully at a full range of energy sources, from oil and gas in our Exclusive Economic Zone to wave and tidal power and ocean thermal energy conversion. Better knowledge of the ocean and seabed is necessary to exploit responsibly the ocean's untapped petroleum and natural gas resources.

Coastal Hazards

This nation must improve its prediction of and response to coastal hazards, both natural and human induced. Hurricanes Hugo and Andrew, two of the strongest hurricanes of the century, devastated parts of the U.S. East Coast. Their impact reinforced the need for better predictive capabilities and a better understanding of coastal storm surges, flooding, erosion, and winds. The exploration for, and production of, petroleum and the transportation of petroleum and chemical products pose risks to the environment when spillage occurs. The movement, effects, and ultimate fates of spilled products must be understood for effective public response. The available information is woefully inadequate, particularly for fragile ecosystems such as coral reefs.

CONTINENTAL MARGINS

Continental margins refer to the region of transition from the land to the deep seafloor, i.e. between continental and oceanic crust. In an active continental margin, the boundary between the continent and the ocean is also a tectonic plate boundary, so there is a lot of geological activity around the margin. The west coast of the United States is an example of an active margin, where the coastline corresponds with the boundary between the Pacific and North America Plates. A passive continental margin occurs where the transition from land to sea is not associated with a plate boundary. The east coast of the United States is a good example; the plate boundary is located along the mid Atlantic ridge, far from the coast. Passive margins are less geologically active. Figure 1 shows an idealized passive margin. When examining this figure, and others like it, note that there is significant vertical exaggeration; the depth scale covers approximately 5000 m, while the horizontal scale extends around 300 km. This makes the features look much steeper than they actually are. The bar at the bottom of Fig. 1 shows what a passive margin would look like without this exaggeration; there is a much more gradual transition to depth.

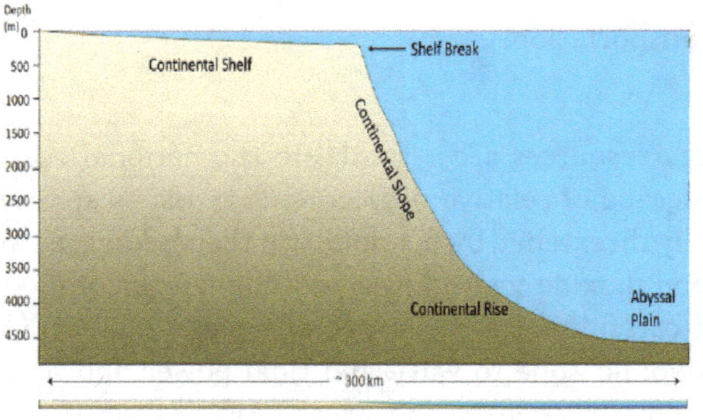

Fig. 1. Features of a passive continental margin

The **continental shelf** is the shallow, flooded edge of the continent. Geologically the shelf is still part of the continental crust, but it is often overlaid with marine sediments. On average, the shelf extends about 80 km from the coast; some margins have very little shelf, while the Siberian Shelf in the Arctic extends roughly 1500 km. The depth of the shelf generally remains below about 150 m, and the floor of the shelf is fairly flat. The flat topography is the result of changes in sea level; throughout history the shelves have been both submerged and exposed, and as sea level rose and fell, wave action, ice sheets, and other erosional processes smoothed out the shelf surface. Wave action and the movement of sediments over the shelf have continued this smoothing process. Continental shelves only make up about 6% of the ocean's surface area, but they are biologically one of the richest parts of the ocean; their shallow depth prevents nutrients from sinking out, and their proximity to the coast provides significant nutrient input. The continental shelf ends at the shelf break, which is the point where the angle of the seafloor begins to get steeper. The shelf break averages about 135 m deep.

After the shelf break, the seafloor takes on a steeper angle (about 4°) as it descends to the deep ocean. This steeper portion of the margin is the continental slope, and it extends from the shelf break down to 3000-5000m. In some parts of the ocean, large submarine canyons have been carved into the continental slope; for example, Monterey Canyon in Monterey Bay, California, is a submarine canyon similar in size to the Grand Canyon! These canyons may be carved out by turbidity currents, which are essentially landslides of sediment, rocks, and other debris down the face of the slope.

At the bottom of the slope is the **continental rise**. This area represents where the continental crust meets the oceanic crust, as the slope begins to level off to become the deep ocean floor. The rise consists of a thick layer of accumulated sediment coming from the continent, so it is difficult to tell where the slope ends and the rise begins.

After the rise comes the **abyssal plain**, or the deep ocean floor, lying between 4500-6000 m. The abyssal plain includes most of the ocean floor, and is the flattest region on Earth. It is flat due to millions of years of sediment accumulation on the bottom, which buries many bottom features (Fig. 2).

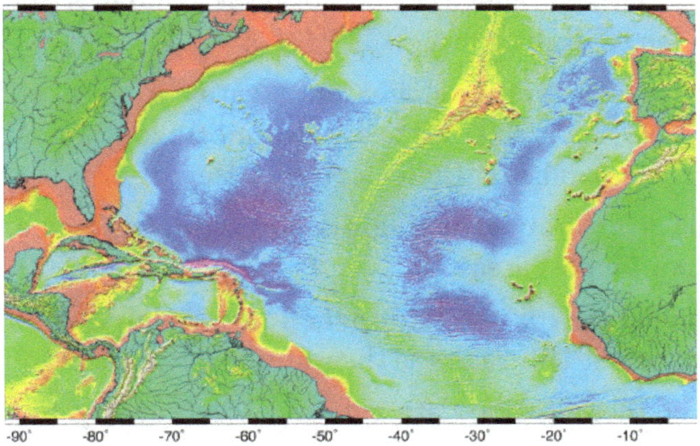

Fig. 2. Topography of the North Atlantic. Shallow continental shelf regions are shown in red, and the abyssal plain is shown in blue. Along the east coast of the United States the continental slope can be seen in green

Passive margins, as described above, have wide shelves, gentle slopes, and a well-developed rise. Since passive margins are not plate boundaries, they experience long periods of relative stability which can lead to the development of these features. Active margins have similar features to passive margins, but the plate boundary affects the properties of the features. Active margins, like the Pacific coast of North America, have narrower shelves, steeper slopes, and little to no rise, particularly in convergent boundaries. Trenches associated with subduction zones act as sediment traps, preventing the accumulation of a continental rise, and keeping sediments off of the abyssal plains.

MARINE PROVINCES

The first major distinction is between the pelagic and benthic zones. The pelagic zone refers to the water column, where swimming and floating organisms live. The benthic zone refers to the bottom, and organisms living on and in the bottom are known as the benthos.

The pelagic zone is divided into two provinces: the **neritic** province corresponds to all of the water from the low tide line to the shelf break, while the **oceanic** province represents all of the other water in the open ocean regions.

The oceanic province is divided into depth zones (Fig. 3):

- 0-200 m is the **epipelagic** zone ("epi" = "upon", as in on top of the pelagic zone). This is the region where enough light penetrates the water to support photosynthesis, so it is also called the **euphotic** or **photic** zone.
- 200-1000 m is the **mesopelagic** zone ("meso" = "middle"). There is some light here, but not enough for photosynthesis, so it is called the **dysphotic zone**, or the **twilight zone**.
- 1000-4000 m is the **bathypelagic** zone ("bathy" = "deep"). There is no light at these depths, so it is referred to as the **aphotic** zone. About 75% of the living space in the ocean lies at these depths or deeper.
- 4000-6000 m is the **abyssopelagic** or **abyssalpelagic** zone, which extends to the seafloor in most areas.
- 6000 m and below is the **hadopelagic** or **hadalpelagic** zone (named for Hades or "hell"). This refers to the water in deep ocean trenches.

Inhabitants of these regions are referred to according to their habitat, for example mesopelagic fish, or epipelagic squid.

The benthic environment is also divided into zones, most of which correspond to the pelagic divisions:

- The **supralittoral** zone lies above the high tide line. Also called the spray zone, it is only submerged during storms or unusually high waves.

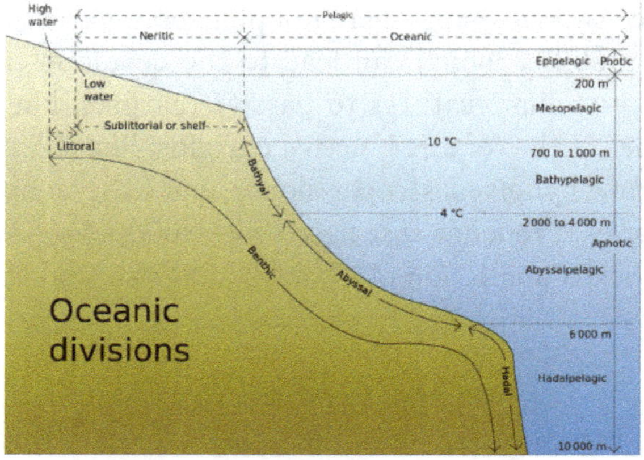

Fig. 3. The major benthic and pelagic oceanic divisions

- The **littoral** zone is the region between the high and low tides. Thus it is also referred to as the **intertidal** zone.

- Below the littoral zone is the **sublittoral** (shelf) zone, extending from the low tide mark to the shelf break, essentially covering the continental shelf.
- The **bathyal** zone extends along the bottom from the shelf break to 4000m, so it generally includes the continental slope and rise.
- The **abyssal** zone is found between 4000-6000 m, including most of the abyssal plains. The abyssal zone represents about 80% of the benthic environment.
- The **hadal** zone includes all benthic regions deeper than 6000 m, such as in the bottom of trenches.

MAPPING THE SEAFLOOR

The previous sections included some information on the depths of the oceans in various places. So how are we able to map the ocean floor to ascertain these vast depths?

To map the ocean floor we need to know the depth at a number of places. The process of measuring the depths is known as **bathymetry**. These measurements were first made through **soundings**, where a weighted line (lead line) was let out by hand until it touched the bottom, and the depth could be recorded from the length of the line. This technique led to the fathom as a unit of depth; as sailors hauled in the sounding line they would stretch it out to cover their arm span. The average arm span of a sailor was about six feet, so one fathom equals six feet, and the sailors could simply count the number of "arm spans" as they pulled in the line.

This technique had a number of drawbacks, and was usually limited to shallower water. It was very time consuming, and only gave depth data for a single point, so many individual soundings were needed to map an area. It could also be error-prone; in deep water it could be difficult to determine when the weight hit the bottom as the weight of the line itself could cause the line to keep sinking, and currents could deflect the line away from vertical, thus overestimating the depth. In later years, winches and heavy steel cables were used for deeper water, but this did not solve all of the problems inherent in the sounding method, and also added the constraint of excessive weight of the equipment.

In the 19th century, a number of modifications were made to this simple design. In 1802 the British clockmaker Edward Massey invented a mechanical device that was attached to the sounding line; as the device sank, a rotor turned a dial which locked in place when the line hit bottom (Fig. 4). The line could then be reeled in and the depth read from the dial. In 1853 American sailor John Mercer Brooke developed a cannonball weight attached to a twine reel. The cannonball was dropped over the side and allowed to free-fall to the bottom; by timing the fall rate (the rate at which the twine unspooled) and noting when the rate changed as the cannonball hit the bottom, the water depth could be calculated. When it hit bottom, the cannonball was released and the line could be hauled back in, bringing with it a sample of mud in the iron bar that held the cannonball, thus confirming that the bottom had been reached.

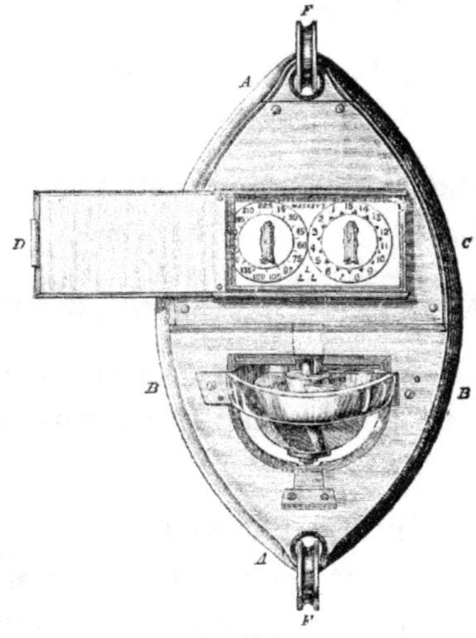

Fig. 4. Massey's sounding machine

After the *Titanic* disaster in 1912, there was an effort to develop better methods of detecting icebergs from a ship. This led to the development of **sonar** (SOund Navigation And Ranging) technology, which was soon applied to mapping bathymetry. A sonar device called an echosounder sends out a pulse of sound, then listens for the returning echo. The timing of the returning echo is used to calculate depth. We know that the speed of sound in water is approximately 1500 m/s. Since the returning echo traveled to the bottom and back, the water depth corresponds to half the time it takes for an echo to return, multiplied by the speed of sound in water (Fig. 5):

depth=1/2*(two-way travel time)*(speed of sound in water)

Echosounders allowed a fast, continuous record of bathymetry under a moving ship. However, they only give the depth directly under a ship's path. Today, high resolution seafloor maps are made through multibeam or side scan sonar, either from a ship or from a towed transmitter (Fig. 6). Multibeam sonar produces a fan-shaped acoustic field allowing a much a wider area (>10 km wide) to be mapped simultaneously.

Large-scale mapping of the ocean floor is also carried out by satellites (originally SEASAT, then GEOSAT, now the Jason satellites) which use radio waves to measure the height of the sea surface (radar altimetry). The sea surface is not flat; gravity causes it to be slightly higher over elevated features on the ocean floor, and slightly lower over trenches and other depressions. Satellites send out radio waves, and similar to an echosounder, can use the returning waves to detect differences in sea surface height down to 3-6 cm (Fig. 7). These differences in sea surface heights allow us to determine the topography under the surface. Unlike the old lead line technology, where hundreds of soundings were necessary to map a small area, each taking an hour or more to complete, the current satellites can map over 90% of Earth's ice-free sea surface every 10 days.

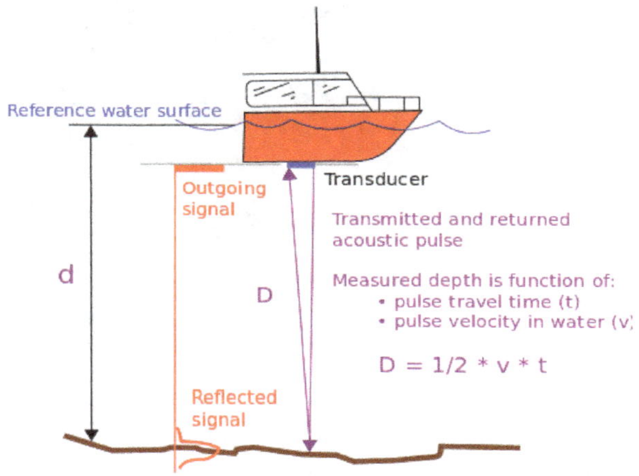

Fig. 5 Measuring depth using an echosounder

Fig. 6. Multibeam sonar (NOAA).

PHYSICAL AND CHEMICAL PROPERTIES OF OCEAN WATERS

The physical and chemical properties of seawater vary according to latitude, depth, nearness to land, and input of fresh water. Approximately 3.5 percent of seawater is composed of dissolved compounds, while the other 96.5 percent is pure water. The chemical composition of seawater reflects such processes as erosion of rock and sediments, volcanic activity, gas exchange with the atmosphere, the metabolic and breakdown products of organisms, and rain. In addition to carbon, the nutrients essential for living organisms include nitrogen and phosphorus, which are minor constituents of seawater and thus are often limiting factors in organic cycles of the ocean. Concentrations of phosphorus and nitrogen are generally low in the photic zone because they are rapidly taken up by marine organisms. The highest concentrations of these nutrients generally are found below 500 metres, a result of the decay of organisms. Other important elements include silicon (used in the skeletons of radiolarians and diatoms) and calcium (essential in the skeletons of many organisms such as fish and corals).

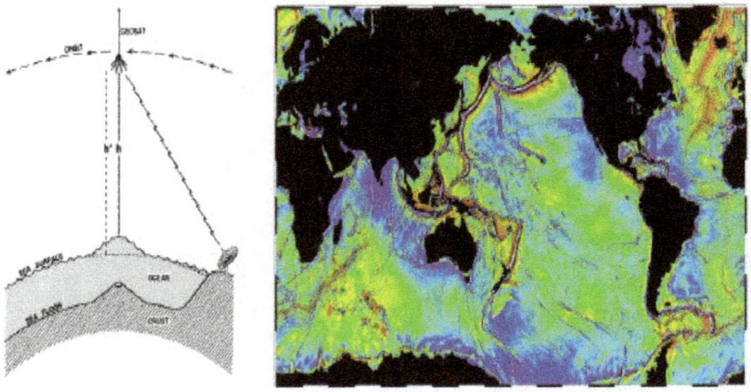

Fig. 7. Radar altimetry (left) and a map of the seafloor produced by radar altimetry satellites (right) (NOAA).

The chemical composition of the atmosphere also affects that of the ocean. For example, carbon dioxide is absorbed by the ocean and oxygen is released to the atmosphere through the activities of marine plants. The dumping of pollutants into the sea also can affect the chemical makeup of the ocean, contrary to earlier assumptions that, for example, toxins could be safely disposed of there.

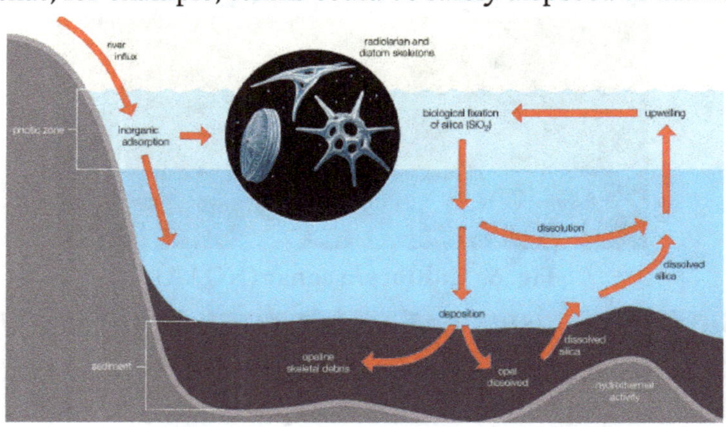

Fig. 8. Silica cycle

The physical and chemical properties of seawater have a great effect on organisms, varying especially with the size of the creature. As an example, seawater is viscous to very small animals (less than 1 millimetre [0.039 inch] long) such as ciliates but not to large marine creatures such as tuna.

Marine organisms have evolved a wide variety of unique physiological and morphological features that allow them to live in the sea. Notothenid fishes in Antarctica are able to inhabit waters as cold as -2° C (28° F) because of proteins in their blood that act as antifreeze. Many organisms are able to achieve neutral buoyancy by secreting gas into internal chambers, as cephalopods do, or into swim bladders, as some fish do; other organisms use lipids, which are less dense than water, to achieve this effect. Some animals, especially those in the aphotic zone, generate light to attract prey. Animals in the disphotic zone such as hatchetfish produce light by means of organs called photophores to break up the silhouette of their bodies and avoid visual detection by predators. Many marine animals can detect vibrations or sound in the water over great distances by means of specialized organs.

Certain fishes have lateral line systems, which they use to detect prey, and whales have a sound-producing organ called a melon with which they communicate. Tolerance to differences in salinity varies greatly: stenohaline organisms have a low tolerance to salinity changes, whereas euryhaline organisms, which are found in areas where river and sea meet (estuaries), are very tolerant of large changes in salinity. Euryhaline organisms are also very tolerant of changes in temperature. Animals that migrate between fresh water and salt water, such as salmon or eels, are capable of controlling their osmotic environment by active pumping or the retention of salts. Body architecture varies greatly in marine waters. The body shape of the cnidarian by-the-wind-sailor (*Velella velella*)—an animal that lives on the surface of the water (pleuston) and sails with the assistance of a modified flotation chamber—contrasts sharply with the sleek, elongated shape of the barracuda.

Ocean currents

The movements of ocean waters are influenced by numerous factors, including the rotation of the Earth (which is responsible for the Coriolis effect), atmospheric circulation patterns that influence surface waters, and temperature and salinity gradients between the tropics and the polar regions (thermohaline circulation). The resultant patterns of circulation range from those that cover great areas, such as the North Subtropical Gyre, which follows a path thousands of kilometres long, to small-scale turbulences of less than one metre.

Marine organisms of all sizes are influenced by these patterns, which can determine the range of a species. For example, krill (*Euphausia superba*) are restricted to the Antarctic Circumpolar Current. Distribution patterns of both large and small pelagic organisms are affected as well. Mainstream currents such as the Gulf Stream and East Australian Current transport larvae great distances. As a result cold temperate coral reefs receive a tropical infusion when fish and invertebrate larvae from the tropics are relocated to high latitudes by these currents. The successful recruitment of eels to Europe depends on the strength of the Gulf Stream to transport them from spawning sites in the Caribbean. Areas where the ocean is affected by nearshore features, such as estuaries, or areas in which there is a vertical salinity gradient (halocline) often exhibit intense biological activity. In these environments, small organisms can become concentrated, providing a rich supply of food for other animals.

Marine biota

Marine biota can be classified broadly into those organisms living in either the pelagic environment (plankton and nekton) or the benthic environment (benthos). Some organisms, however, are benthic in one stage of life and pelagic in another. Producers that synthesize organic molecules exist in both environments. Single-celled or multicelled plankton with photosynthetic pigments are the producers of the photic zone in the pelagic environment. Typical benthic producers are microalgae (e.g., diatoms), macroalgae (e.g., the kelp *Macrocystis pyrifera*), or sea grass (e.g., *Zostera*).

Plankton

Plankton are the numerous, primarily microscopic inhabitants of the pelagic environment. They are critical components of food chains in all marine environments because they provide nutrition for the nekton (e.g., crustaceans, fish, and squid) and benthos (e.g., sea squirts and sponges). They also exert a global effect on the biosphere because the balance of components of the Earth's atmosphere depends to a great extent on the photosynthetic activities of some plankton.

The term *plankton* is derived from the Greek *planktos*, meaning wandering or drifting, an apt description of the way most plankton spend their existence, floating with the ocean's currents. Not all plankton, however, are unable to control their movements, and many forms depend on self-directed motions for their survival.

Plankton range in size from tiny microbes (1 micrometre [0.000039 inch] or less) to jellyfish whose gelatinous bell can reach up to 2 metres in width and whose tentacles can extend over 15 metres. However, most planktonic organisms, called plankters, are less than 1 millimetre (0.039 inch) long. These microbes thrive on nutrients in seawater and are often photosynthetic. The plankton include a wide variety of organisms such as algae, bacteria, protozoans, the larvae of some animals, and crustaceans. A large proportion of the plankton are protists—i.e., eukaryotic, predominantly single-celled organisms. Plankton can be broadly divided into phytoplankton, which are plants or plantlike protists; zooplankton, which are animals or animal-like protists; and microbes such as bacteria. Phytoplankton carry out photosynthesis and are the producers of the marine community; zooplankton are the heterotrophic consumers.

Diatoms and dinoflagellates (approximate range between 15 and 1000 micrometres in length) are two highly diverse groups of photosynthetic protists that are important components of the plankton. Diatoms are the most abundant phytoplankton. While many dinoflagellates carry out photosynthesis, some also consume bacteria or algae. Other important groups of protists include flagellates, foraminiferans, radiolarians, acantharians, and ciliates. Many of these protists are important consumers and a food source for zooplankton.

Zooplankton, which are greater than 0.05 millimetre in size, are divided into two general categories: meroplankton, which spend only a part of their life cycle—usually the larval or juvenile stage—as plankton, and holoplankton, which exist as plankton all their lives. Many larval meroplankton in coastal, oceanic, and even freshwater environments (including sea urchins, intertidal snails, and crabs, lobsters, and fish) bear little or no resemblance to their adult forms. These larvae may exhibit features unique to the larval stage, such as the spectacular spiny armour on the larvae of certain crustaceans (e.g., *Squilla*), probably used to ward off predators.

CHAPTER

INTRODUCTION TO PLANET "EARTH" 2

Earth, our home, is the third planet from the sun. While scientists continue to hunt for clues of life beyond Earth, our home planet remains the only place in the universe where we've ever identified living organisms.

Earth is the fifth-largest planet in the solar system. It's smaller than the four gas giants - Jupiter, Saturn, Uranus and Neptune - but larger than the three other rocky planets, Mercury, Mars and Venus.

Earth has a diameter of roughly 8000 miles (13,000 kilometers) and is mostly round because gravity generally pulls matter into a ball. But the spin of our home planet causes it to be squashed at its poles and swollen at the equator, making the true shape of the Earth an "oblate spheroid."

Our planet is unique for many reasons, but its available water and oxygen are two defining features. Water covers roughly 71% of Earth's surface, with most of that water located in our planet's oceans. About a fifth of Earth's atmosphere consists of oxygen, produced by plants.

PLANET EARTH'S ORBIT AROUND THE SUN

While Earth orbits the sun, the planet is simultaneously spinning around an imaginary line called an axis that runs through the core, from the North Pole to the South Pole. It takes Earth 23.934 hours to complete a rotation on its axis and 365.26 days to complete an orbit around the sun - our days and years on Earth are defined by these gyrations.

Earth's axis of rotation is tilted in relation to the ecliptic plane, an imaginary surface through the planet's orbit around the sun. This means the Northern and Southern hemispheres will sometimes point toward or away from the sun depending on the time of year, and this changes the amount of light the hemispheres receive, resulting in the changing seasons.

Earth happens to orbit the sun within the so-called "Goldilocks zone," where temperatures are just right to maintain liquid water on our planet's surface. Earth's orbit is not a perfect circle, but rather a slightly oval-shaped ellipse, similar to the orbits of all the other planets in our solar system. Our planet is a bit closer to the sun in early January and farther away in July, although this proximity has a much smaller effect on the temperatures we experience on the planet's surface than does the tilt of Earth's axis.

Statistics about Earth's orbit, according to NASA:

- Average distance from the sun: 92,956,050 miles (149,598,262 km)
- Perihelion (closest approach to the sun): 91,402,640 miles (147,098,291 km)
- Aphelion (farthest distance from the sun): 94,509,460 miles (152,098,233 km)
- Length of solar day (single rotation on its axis): 23.934 hours
- Length of year (single revolution around the sun): 365.26 days
- Equatorial inclination to orbit: 23.4393 degrees

EARTH'S FORMATION AND DEVELOPMENT

Scientists think Earth was formed at roughly the same time as the sun and other planets some 4.6 billion years ago when the solar system coalesced from a giant, rotating cloud of gas and dust known as the solar nebula. As the nebula collapsed under the force of its own gravity, it spun faster and flattened into a disk. Most of the material in that disk was then pulled toward the center to form the sun.

Fig. 1. Planets in the solar system follow different orbit lines around the sun.

Other particles within the disk collided and stuck together to form ever-larger bodies, including Earth. Scientists think Earth started off as a waterless mass of rock.

"It was thought that because of these asteroids and comets flying around colliding with Earth, conditions on early Earth may have been hellish," Simone Marchi, a planetary scientist at the Southwest Research Institute in Boulder, Colorado, previously told Space.com.

However, analyses of minerals trapped within ancient microscopic crystals suggest that there was liquid water already present on Earth during its first 500 million years, Marchi said.

Radioactive materials in the rock and increasing pressure deep within the Earth generated enough heat to melt the planet's interior, causing some chemicals to rise to the surface and form water, while others became the gases of the atmosphere. Recent evidence suggests that Earth's crust and oceans may have formed within about 200 million years after the planet took shape.

EARTH'S ATMOSPHERE

Earth's atmosphere is roughly 78 percent nitrogen and 21 percent oxygen, with trace amounts of water, argon, carbon dioxide and other gases. No other planet in the solar system has an atmosphere loaded with free oxygen, which is vital to one of the other unique features of Earth: life.

Air surrounds Earth and becomes thinner farther from the surface. Roughly 100 miles (160 km) above Earth, the air is so thin that satellites can zip through the atmosphere with little resistance. Still, traces of atmosphere can be found as high as 370 miles (600 km) above the planet's surface.

The lowest layer of the atmosphere is known as the troposphere, which is constantly in motion and why we have weather. Sunlight heats the planet's surface, causing warm air to rise into the troposphere. This air expands and cools as air pressure decreases, and because this cool air is denser than its surroundings, it then sinks and gets warmed by the Earth again.

Above the troposphere, some 30 miles (48 km) above the Earth's surface, is the stratosphere. The still air of the stratosphere contains the ozone layer, which was created when ultraviolet light caused trios of oxygen atoms to bind together into ozone molecules. Ozone prevents most of the sun's harmful ultraviolet radiation from reaching Earth's surface, where it can damage and mutate life.

Water vapor, carbon dioxide and other gases in the atmosphere trap heat from the sun, warming Earth. Without this so-called "greenhouse effect," Earth would probably be too cold for life to exist, although a runaway greenhouse effect led to the hellish conditions of Venus' current surface.

Earth-orbiting satellites have shown that the upper atmosphere actually expands during the day and contracts at night due to heating and cooling.

EARTH'S CHEMICAL COMPOSITION

Oxygen is the most abundant element in rocks in Earth's crust, composing roughly 47 percent of the weight of all rock. The second most abundant element is silicon, at 27 percent, followed by aluminum, at 8 percent; iron, at 5%; calcium, at 4%; and sodium, potassium and magnesium, at about 2% each.

Earth's core consists mostly of iron and nickel and potentially smaller amounts of lighter elements, such as sulfur and oxygen. The mantle is made of iron and magnesium-rich silicate rocks (the combination of silicon and oxygen is known as silica, and minerals that contain silica are known as silicate minerals).

EARTH FROM SPACE

Astronauts and scientists have learned a lot about our planet by leaving it. From 240 miles (408 kilometers) away, aboard the International Space Station (ISS), astronauts are able to observe the thin, fragile atmosphere of Earth.

Fig. 2. Earth, pictured from the International Space Station

Meanwhile, satellites orbiting Earth can track the planet's responses to changes —caused naturally and by humans– on a greater scale. Satellites have been launched to detect changes in the hole in the ozone layer, monitor cloud coverage and weather patterns and manage humans' use of Earth's resources.

LIFE ON EARTH

Earth is the only planet in the universe known to possess life. The planet boasts several million described species, living in habitats ranging from the bottom of the deepest ocean to a few miles up into the atmosphere. Researchers think far more species remain that have yet to be described to science.

Researchers suspect that other candidates for hosting life in our solar system — such as Saturn's moon Titan or Jupiter's moon Europa — could house primitive living creatures. Scientists have yet to precisely nail down exactly how our primitive ancestors first showed up on Earth, although most believe that a chemical soup on the planet gave rise to the building blocks of living organisms.

Another theory suggests that life first evolved on the nearby planet Mars, which could once have been habitable, then traveled to Earth on meteorites hurled from the Red Planet by impacts from other space rocks.

"It's lucky that we ended up here, nevertheless, as certainly Earth has been the better of the two planets for sustaining life," biochemist Steven Benner, of the Westheimer Institute for Science and Technology in Florida, told Space.com. "If our hypothetical Martian ancestors had remained on Mars, there might not have been a story to tell."

HOW ARE EARTH'S OCEANS UNIQUE?

We often think of Earth in terms of its land area, but in reality 71% of the Earth's surface is covered by oceans, while only 29% is land. Oceans cover an area of 139 million miles^2 or 361 million km^2, and contain a volume of about 1.37 billion km^3 of water. All of this water is not distributed equally over the Earth; 61% of the Northern Hemisphere is covered by oceans, while in the Southern Hemisphere the oceans cover 81% c

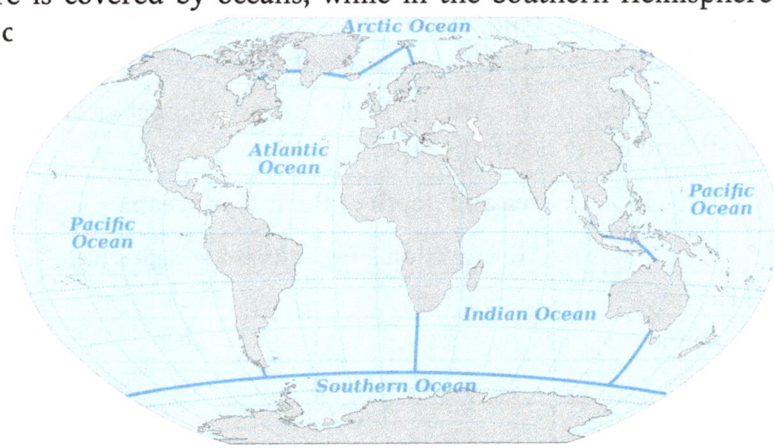

Fig. 3. Map of the world oceans

Various sources differ in the number of recognized ocean basins. Historically the major oceans were recognized as the Pacific, Atlantic, Indian, and Arctic Oceans. More recently, the Southern Ocean has been recognized as fifth named ocean, comprising all of the water from the coast of Antarctica to 60° S (Figure 1). In 2000 these boundaries were submitted to the International Hydrographic Organization for official recognition, but several countries do not recognize it as a separate ocean, but rather as the southern extension of the other major oceans. The Southern Ocean has its own unique characteristics, so for the purposes of this book we will include it as a separate ocean.

The oceans account for vast amounts of water, containing 97% of the water on Earth's surface, with over half of the water in the Pacific alone (Table 1).

Table 1. Percentage of Earth's water in various locations

Pacific	52%
Atlantic	25%
Indian	20%
Ice	2%
Ground water	0.6%
Atmosphere, lakes & rivers	0.01%

The average depth of the world ocean is about 3800m (12,500 ft), which is about four times deeper than the average land elevation is high (840m or 2800 ft). In fact Mt. Everest, the highest point on land, is 8848m (29,028 ft) high, while the deepest part of the ocean, the Challenger Deep of the Marianas Trench is approximately 10,920m (36,200 ft) deep. So you could submerge Mt. Everest in the Marianas Trench and it would still be covered by over 2 km of water! Because there is so much more water on Earth than there is land, if you could smooth out the land elevation the entire Earth would still be covered by water about 2700 m deep.

Of the major ocean basins, the Pacific is the largest (almost as large as all of the others combined), and is the deepest (Table 2).

Table 2 Area and depth of the major oceans

	Area (million km^2)	Average depth (m)
Pacific	166	4282
Atlantic	87	3926
Indian	73	3963
Arctic	14	1205
Southern	20	4000

EARTH'S OCEANS ARE UNIQUE IN THE UNIVERSE

Earth is the only known planet or moon to have large bodies of liquid water on its surface. Our planet lies in the "Goldilocks" zone — not too hot, not too cold and with enough atmospheric pressure to prevent liquid surface water from evaporating into space.

Although we don't yet know of any other planets or moons with liquid water oceans, it's likely that they do exist and we just haven't found them. In our own solar system, there is growing evidence that the planet Mars may have liquid water not on the surface but underground.

There is also strong evidence that liquid oceans may be hidden beneath the thick icy surfaces of three of Jupiter's moons (Europa, Callisto and Ganymede) and two of Saturn's moons (Titan and Enceladus).

But nobody knows exactly how many different species live in this environment. There is no data for around 20 per cent of the ocean's volume.

The Census of Marine Life, a 10-year international project to identify life in our oceans, found nearly 250,000 species. But scientists believe a least a million species of marine life could be out there, and that's not counting the tens or even hundreds of millions of kinds of microbes that make up the majority of marine life.

What we do know is that ocean life survives in the most extreme environments. Scientists have found life that can survive in temperatures that melt lead, where seawater freezes into ice, or there's no light or oxygen.

In fact, the dark ocean zone between 1000 and 5000 metres known as the abyssal zone has a far greater range of marine life than we once thought.

The deep waters around Australia's Great Barrier Reef are home to a stunning array of life such as the deep sea jellyfish atolla, which was discovered by Australian scientists in 2006.

Water takes around 1000 years to travel all the way around the whole globe

The oceans not only have waves, tides and surface currents — they also have a constantly moving system of deep-ocean circulation driven by temperature and salinity.

Known as the global ocean conveyor belt or thermohaline current (thermo = temperature, haline = salinity), this deep ocean current gets one of its "starts" in the polar region near Norway.

As sea ice forms, the water left behind becomes saltier and denser and begins to sink, making room for warmer and less dense incoming surface water, which in turn eventually becomes cold and salty enough to sink.

The cold dense water flows along the ocean bottom all the way from the northern hemisphere to the Southern Ocean where it merges with more cold dense water from Antarctica and is swept into the Indian and Pacific Oceans.

Eventually it mixes with warmer water and rises to the surface before finding its way back to the Atlantic. It can take 1000 years to complete this cycle.

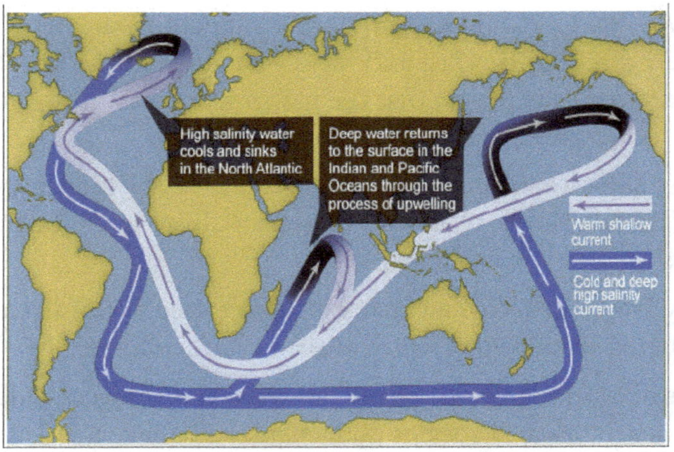

Fig. 4. The Global Ocean Conveyor Belt

Half of all the oxygen we breathe is produced in the ocean

Some of this oxygen is produced by sea weeds and sea grasses, but the vast majority of the oxygen is produced by phytoplankton, microscopic single celled organisms that have the ability to photosynthesise. These tiny creatures live in the surface layer of the ocean (and in lakes and rivers) and form the very base of the aquatic food chain.

During photosynthesis, phytoplankton remove carbon dioxide from sea water and release oxygen. The carbon becomes part of their bodies.

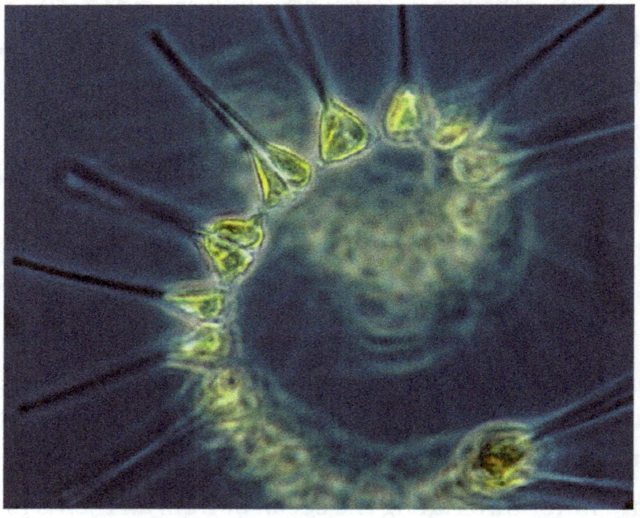

Fig. 5. Phytoplankton - the foundation of the oceanic food chain

The oceans may seem vast, but they're small compared to the overall size of our planet

About 71 per cent of the Earth's surface is covered by water, but the average depth of the oceans is only about 4.2 kilometres. The deepest point is in the Marianas Trench in the Pacific Ocean, around 11 kilometres below sea level.

To get some idea of how the average depth of the oceans compares to the scale of the Earth, imagine if the world's tallest building (the 829-metre Burj Dubai) was the distance from the centre of the Earth to the surface. The average depth of the oceans would be the equivalent of just 10 centimetres at the very top. Mount Everest would be another 21 centimetres above that.

All of the water on Earth (the oceans, which account for 96.5 per cent, plus groundwater, rivers, lakes, snow and ice) would fit into a sphere about 1,385 kilometres in diameter, according to the US Geological Survey. The Earth's diameter is roughly nine times larger (12,700 kilometres).

Rogue waves really do exist

For centuries maritime folklore has had tales of gigantic waves that would appear without warning in mid-ocean and sink ships, even in good weather. These rogue waves are not necessarily the biggest waves ever found at sea, but isolated waves way bigger than what any crew might expect in a given sea state.

By definition, rogue waves are ones whose height is more than twice that of the prevailing conditions.

The largest scientifically measured rogue waves were encountered in February 2000, by the British oceanographic research vessel, the *RRS Discovery*. West of Scotland it recorded individual waves of up to 29.1 metres, almost the height of a 10 storey building.

The causes of these waves are not completely understood, however they seem to happen more often when a strong ocean current runs counter to the direction of the waves.

Oceans hold around 50 times more carbon than the atmosphere

Cold water can dissolve much more CO_2 than warm water, so the cold polar regions are net absorbers of CO_2. But as the cold water finds its way to warmer tropical areas, the oceans release CO_2 back into the atmosphere. The equatorial Pacific is thought to be the biggest single natural source of CO_2 in the atmosphere. Most of this carbon is exchanged with the atmosphere on a timescale of several hundred years.

Prior to the industrial revolution, the uptake and release of CO_2 on land and ocean was in a dynamic equilibrium. Since then, the oceans are thought to have absorbed about half of the carbon dioxide released from the burning of fossil fuels, with the rest remaining in the atmosphere.

Fig. 6. The quick carbon cycle

The great garbage patch is more like a plastic soup

The existence of theThe Great Pacific Garbage 'patch' in the North Pacific Ocean was confirmed in 1997. It lies at the centre of a large rotating ocean current - or gyre. Since then, a soup of plastic pollution has been found at the centre of all the world's major gyres - in the South Pacific, the North and South Atlantic and in the Indian Ocean.

Plastic trash takes a very long time to bio-degrade in the oceans. The actions of sunlight and wave motion tend to break plastic objects into smaller and smaller pieces until they eventually become smaller than a grain of sand. That's why there is no obvious floating island of plastic debris, but rather a fine soup of plastic particles floating in the water column.

These fine particles are now thought to act like sponges, concentrating pollutants such as PCBs, DDT and PAHs. When the particles are ingested by filter feeding sea creatures, they enter the food chain and ultimately into fish destined for human consumption.

They also harbour unique colonies of microbes, but it is too early to say what impact this emerging 'plastisphere' will have on marine ecological environments.

Plastics only came into widespread use after 1945, but can already be found in every part of the marine environment from the surface to the seafloor.

The Atlantic Ocean is getting bigger and the Pacific Ocean is getting smaller

Twenty million years ago there was no Atlantic Ocean. But then, thanks to plate tectonics, the South American and North American continents were separated by a rift valley that eventually turned into the Atlantic Ocean. The complementary shapes of the South American and African continents have long been noticed, but it wasn't until the 1960s that the theory of plate tectonics became accepted as the explanation.

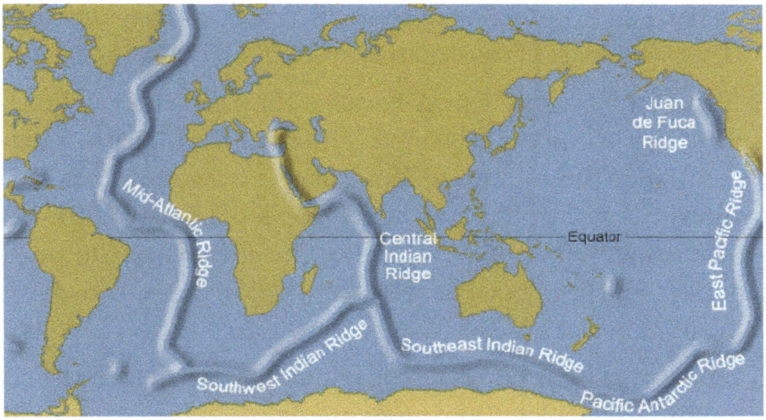

Fig. 7. Mid-ocean ridges

These days, the Atlantic Ocean is growing at a rate of five centimetres per year, as new sea floor is created by volcanic activity along its mid-ocean ridge.

On the other hand, the much older Pacific Ocean is currently estimated to be shrinking by two to three centimetres each year. Again, this comes down to plate tectonics because the Pacific Ocean has subduction zones on three sides — where the Pacific plate submerges beneath other plates.

The Pacific's "Ring of Fire" broadly maps the areas where this subduction occurs and accounts for around 90 per cent of the world's earthquakes.

The most remote point in the oceans is called Point Nemo

The part of the ocean farthest from land lies in the South Pacific and is known as Point Nemo or 'The Pole of Inaccessibility'.

Point Nemo isn't an actual 'point' as (obviously) there's no land there — it's simply a spot in the ocean that happens to be 2,688 kilometres from the nearest land — Ducie Island, part of the Pitcairn Islands, to the north; Motu Nui, one of the Easter Islands, to the northeast; and Maher Island, part of Antarctica, to the south.

The name 'Nemo' comes from Jules Verne's Captain Nemo of *Twenty Thousand Leagues under the Sea* — not the film *Finding Nemo*. If you go to Google Earth and zoom in to 48°52.62 S 123°23.62 W you'll find it thoughtfully marked.

HOW WAS EARLY EXPLORATION OF THE OCEANS ACHIEVED?

Oceans cover 70 percent of the Earth's surface, yet even today their depths remain largely unexplored. Scientists estimate between 90 and 95 percent of the deep sea remains a mystery. The deep sea is truly the planet's final frontier.

The term "deep sea" doesn't have the same meaning to everyone. To fishermen, the deep sea is any part of the ocean beyond the relatively shallow continental shelf. To scientists, the deep sea is the lowest part of the ocean, below the thermocline (the layer where heating and cooling from sunlight ceases to have an effect) and above the sea floor. This is the part of the ocean deeper than 1,000 fathoms or 1,800 meters.

It's difficult to explore the depths because they are eternally dark, extremely cold (between 0°C and 3°C below 3000 meters), and under high pressure (15750 psi or over 1000 times higher than standard atmospheric pressure at sea level). From the time of Pliny until the end of the 19th century, people believed the deep sea was a lifeless wasteland. Modern scientists recognize the deep sea as the largest habitat on the planet. Special tools have been developed to explore this cold, dark, pressurized environment.

Deep sea exploration is a multi-disciplinary endeavor that includes oceanography, biology, geography, archaeology, and engineering.

A Brief History of Deep Sea Exploration

The history of deep sea exploration begins relatively recently, mainly because advanced technology is needed to explore the depths. Some milestones include:

1521: Ferdinand Magellan attempts to measure the depth of the Pacific Ocean. He uses a 2400-foot weighted line, but does not touch bottom.

1818: Sir John Ross catches worms and jellyfish at a depth of approximately 2,000 meters (6550 feet), offering the first evidence of deep sea life.

1842: Despite Ross' discovery, Edward Forbes proposes the Abyssus Theory, which states biodiversity decreases with death and that life cannot exist deeper than 550 meters (1800 feet).

1850: Michael Sars refutes the Abyssus Theory by discovering a rich ecosystem at 800 meters (2600 feet).

1872-1876: The HMS *Challenger*, led by Charles Wyville Thomson, conducts the first deep sea exploration expedition. *Challenger*'s team discovers many new species uniquely adapted to life near the sea floor.

1930: William Beebe and Otis Barton become the first humans to visit the deep sea. Within their steel Bathysphere, they observe shrimp and jellyfish.

1934: Otis Barton sets a new human diving record, reaching 1370 meters (.85 miles).

1956: Jacques-Yves Cousteu and his team aboard the *Calypso* release the first full-color, full-length documentary, *Le Monde du silence* (*The Silent World*), showing people everywhere the beauty and life of the deep sea.

1960: Jacques Piccard and Don Walsh, with the deep sea vessel *Trieste*, descend to the bottom of the Challenger Deep in the Mariana Trench (10,740 meters/6.67 miles). They observe fish and other organisms. Fish were not thought to inhabit such deep water.

1977: Ecosystems around hydrothermal vents are discovered. These ecosystems use chemical energy, rather than solar energy.

1995: Geosat satellite radar data is declassified, allowing for global mapping of the sea floor.

2012: James Cameron, with the vessel *Deepsea Challenger*, completes the first solo dive to the bottom of the Challenger Deep.

Modern studies expand our knowledge of the geography and biodiversity of the deep sea. The *Nautilus* exploration vehicle and NOAA's *Okeanus Explorer* continue to discovery new species, unravel man's effects on the pelagic environment, and explore wrecks and artifacts deep beneath the sea surface. The Integrated Ocean Drilling Program (IODP) *Chikyu* analyzes sediments from the Earth's crust and may become the first ship to drill into the Earth's mantle.

Instrumentation and Technology

Like space exploration, deep sea exploration requires new instruments and technology. While space is a cold vacuum, the ocean depths are cold, but highly pressurized. The saltwater is corrosive and conductive. It's very dark.

Finding the Bottom

In the 8th century, Vikings dropped lead weights attached to ropes to measure water depth. Beginning in the 19th century, researchers used wire rather than rope to take sounding measurements. In the modern era, acoustic depth measurements are the norm. Basically, these devices produce a loud sound and listen for echoes to gauge distance.

Human Exploration

Once people knew where the sea floor was, they wanted to visit and examine it. Science has progressed way beyond the diving bell, a barrel containing air that could be lowered into the water. The first submarine was built by Cornelius Drebbel in 1623.

The first underwater breathing apparatus was patented by Benoit Rouquarol and Auguste Denayrouse in 1865. Jacques Cousteau and Emile Gagnan developed the Aqualung, which was the first true "Scuba" (Self Contained Underwater Breathing Apparatus) system. In 1964, Alvin was tested. Alvin was built by General Mills and operated by the US Navy and Woods Hole Oceanographic Institution.

Alvin allowed three people to remain underwater for as long as nine hours and as deep as 14800 feet. Modern submarines can travel as deep as 20000 feet.

Robotic Exploration

While humans have visited the bottom of the Mariana Trench, the trips were expensive and only allowed limited exploration. Modern exploration relies on robotic systems.

Remotely operated vehicles (ROVs) are tethered vehicles that are controlled by researchers on a ship. ROVs typically carry cameras, manipulator arms, sonar equipment, and sample containers.

Autonomous underwater vehicles (AUVs) operate without human control. These vehicles generate maps, measure temperature and chemicals, and take photographs. Some vehicles, such as the *Nereus*, act as either a ROV or AUV.

Instrumentation

Humans and robots visit locations but don't remain long enough to collect measurements over time. Undersea instruments monitor whale songs, plankton density, temperature, acidity, oxygenation, and various chemical concentrations. These sensors may be attached to profiling buoys, which drift freely at a depth of about 1000 meters. Anchored observatories house instruments on the seafloor. For example, the Monterey Accelerated Research System (MARS) rests on the floor of the Pacific Ocean at 980 meters to monitor seismic faults.

Deep Sea Exploration Fast Facts

- The deepest part of the Earth's oceans is the Challenger Deep in the Mariana Trench, at 10,994 meters (36,070 feet or nearly 7 miles) below sea level.
- Three people have visited the depths of the Challenger Deep. Film director James Cameron reached a record depth of 35,756 feet in a solo submersible dive in 2012.
- Mount Everest would fit inside the Mariana Trench, with over a mile of extra space above it.
- Using bomb sounding (throwing TNT into a trench and recording the echo), scientists have found the Mariana Trench, Kermadec, Kuril-Kamchatka, Philippine, and Tonga trenches all exceed 10000 meters in depth.
- While human exploration still occurs, most modern discoveries are made using data from robots and sensors.

Looking Ahead

- Man's curiosity surrounding one subject can spark tangents that prove to be more widely useful as well. The development of sounding equipment and sonar not only allows scientists to map out the ocean floor, but it also allows for the detection of objects in the water. This technology allows submarines to navigate the dark deep waters. In turn, submarines themselves not only provide transportation and visuals of the deep sea, but also are tactical tools. New and improved diving equipment allows humans to experience the beautiful corals in the ocean for themselves and to uncover mysteries of sunken ships that were seemingly lost to history.

- Only five percent of the ocean has been explored to date. As time moves on, the technology will only improve, and, eventually, 100 percent of the ocean will be covered. What will humans find next in the deep blue sea?

WHAT FIELDS OF SCIENCE DOES OCEANOGRAPHY INCLUDE?

Oceanography, also called **oceanology** or **marine science** is the branch of physical geography that studies the Earth's oceans and seas. Oceanographers study a wide range of topics such as plate tectonics to ocean currents to marine organisms. **Oceanography** includes the branches of science that deal with the physical and biological properties, and observable phenomena of the world **oceans** and **seas.** This oceanography course covers many aspects associated with other disciplines including physical geography, geology (including earth history and astronomy), chemistry, meteorology, biology and ecology. Perhaps most important, human interactions, include general history, exploration, exploitation, and some of the many environmental factors affecting our modern global civilization.

Branches

The study of oceanography has four main branches:

Marine biology or biological oceanography

- the study of the plants and animals (biota) of the oceans and their ecological interaction.

Chemical oceanography

- the study of the chemistry of the ocean;

Marine geology or geological oceanography

- the study of the geology of the ocean floor including plate tectonics; and

Physical oceanography

- studies the ocean's physical attributes including temperature-salinity structure, waves, tides and currents).

Metorological oceanography

- studies the ocean-atmosphere interactions

These branches reflect the fact that many oceanographers are first trained in the exact sciences and then focus on applying their interdisciplinary knowledge, skills and abilities to oceanography.

ORIGIN OF EARTH AND THE SOLAR SYSTEM

According to the **Big Bang theory**, the universe blinked violently into existence 13.77 billion years ago (Fig. 8). The Big Bang is often described as an explosion, but imagining it as an enormous fireball isn't accurate. The Big Bang involved a sudden expansion of matter, energy, and space from a single point. The kind of Hollywood explosion that might come to mind involves expansion of matter and energy *within* space, but during the big bang, space *itself* was created.

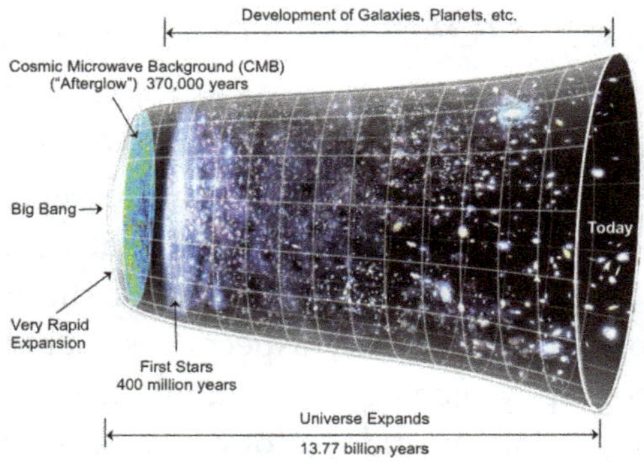

Fig. 8. The Big Bang and development of the universe

At the start of the Big Bang, the universe was too hot and dense to be anything but a sizzle of particles smaller than atoms, but as it expanded, it also cooled. Eventually some of the particles collided and stuck together. Those collisions produced hydrogen and helium, the most common elements in the universe, along with a small amount of lithium. Gravity caused clouds of these early elements to coalesce into stars, and it was inside these stars that heavier elements were formed

Our solar system began to form around 5 billion years ago, roughly 8.7 billion years after the Big Bang. A **solar system** consists of a collection of objects orbiting one or more central stars. All solar systems start out the same way. They begin in a cloud of gas and dust called a **nebula.** Nebulae are some of the most beautiful objects that have been photographed in space, with vibrant colors from the gases and dust they contain, and brilliant twinkling from the many stars that have formed within them. The gas consists largely of hydrogen and helium, and the dust consists of tiny mineral grains, ice crystals, and organic particles.

A solar system begins to form when a small patch within a nebula (small by the standards of the universe, that is) begins to collapse upon itself. Exactly how this starts isn't clear, although it might be triggered by the violent behavior of nearby stars as they progress through their life cycles. Energy and matter released by these stars might compress the gas and dust in nearby neighborhoods within the nebula. Once it is triggered, the collapse of gas and dust within that patch continues for two reasons.

One of those reasons is that gravitational force pulls gas molecules and dust particles together. But early in the process, those particles are very small, so the gravitational force between them isn't strong. So how do they come together? The answer is that dust first accumulates in loose clumps for the same reason dust bunnies form under your bed: static electricity. As the small patch within a nebula condenses, a star begins to form from material drawn into the center of the patch, and the remaining dust and gas settle into a disk that rotates around the star. The disk is where planets eventually form, so it's called a protoplanetary disk. Notice the dark rings in the protoplanetary disk.

These are gaps where planets are beginning to form. The rings are there because incipient planets are beginning to collect the dust and gas in their orbits. There is an analogy for this in our own solar system, because the dark rings are akin to the gaps in the rings of Saturn, where moons can be found.

In general, planets can be classified into three categories based on what they are made of (Fig. 9). Terrestrial planets are those planets like Earth, Mercury, Venus, and Mars that have a core of metal surrounded by rock.

Jovian planets (also called gas giants) are those planets like Jupiter and Saturn that consist predominantly of hydrogen and helium. Ice giants are planets such as Uranus and Neptune that consist largely of water ice, methane (CH_4) ice, and ammonia (NH_3) ice, and have rocky cores. Often, the ice giant planets Uranus and Neptune are grouped with Jupiter and Saturn as gas giants; however, Uranus and Neptune are very different from Jupiter and Saturn.

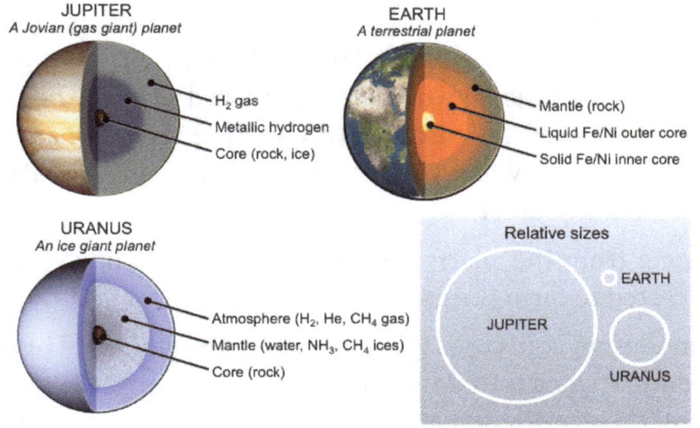

Fig. 9. Three types of planets. Jovian (or gas giant) planets such as Jupiter consist mostly of hydrogen and helium. They are the largest of the three types. Ice giant planets such as Uranus are the next largest. They contain water, ammonia, and methane ice. Terrestrial planets such as Earth are the smallest, and they have metal cores covered by rocky mantles.

These three types of planets are not mixed together randomly within our solar system. Instead they occur in a systematic way, with terrestrial planets closest to the sun, followed by the Jovian planets and then the ice giants. Part of the reason for this arrangement is the frost line (also referred to as the snow line).

The frost line separated the inner part of the protoplanetary disk closer to the sun, where it was too hot to permit anything but silicate minerals and metal to crystallize, from the outer part of the disk farther from the Sun, where it was cool enough to allow ice to form. As a result, the objects that formed in the inner part of the protoplanetary disk consist largely of rock and metal, while the objects that formed in the outer part consist largely of gas and ice. The young sun also blasted the solar system with raging solar winds (winds made up of energetic particles), which helped to drive lighter molecules toward the outer part of the protoplanetary disk.

The objects in our solar system formed by accretion. Early in this process, mineral and rock particles collected in fluffy clumps because of static electricity. As the mass of the clumps increased, gravity became more important, pulling material from farther away and growing these solid masses into larger and larger bodies. Eventually the mass of the objects became large enough that their gravity was strong enough to hang onto gas molecules, because gas molecules are very light.

Our Earth formed though this process of accretion about 4.6 billion years ago. The early Earth was very hot and had a molten, fluid composition, with lost of geological and volcanic activity on the surface. The Earth's heat came from a variety of processes:

- Heat came from the decay of radioactive elements within the Earth, specifically the decay of 235U, 238U, 40K, and 232Th, which are primarily present in the mantle. The total heat produced that way has been decreasing over time (because these isotopes are getting used up), and is now roughly 25% of what it was when Earth formed. This means that Earth's interior is slowly becoming cooler.
- Heat came from the thermal energy already contained within the objects that accreted to form the Earth.
- Heat came from collisions. When objects hit Earth, some of the energy from their motion went into deforming Earth, and some of it was transformed into heat. (The very worst collision that Earth experienced was with a planet named Theia, which was approximately the size of Mars. Not long after Earth formed, Theia struck Earth. When Theia slammed into Earth, Theia's metal core merged with Earth's core, and debris from the outer silicate layers was cast into space, forming a ring of rubble around Earth. The material within the ring coalesced into a new body in orbit around Earth, giving us our moon.

Remarkably, the debris may have coalesced in 10 years or fewer! This scenario for the formation of the moon is called the giant impact hypothesis.)

- As Earth became larger, its gravitational force became stronger. This increased Earth's ability to draw objects to it, but it also caused the material making Earth to be compressed, rather like Earth giving itself a giant gravitational hug. Compression causes materials to heat up.

Heating had a very important consequence for Earth's structure. As Earth grew, it collected a mixture of silicate mineral grains as well as iron and nickel. These materials were scattered throughout Earth. That changed when Earth began to heat up: it got so hot that both the silicate minerals and the metals melted. The metal melt was much denser than the silicate mineral melt, so the metal melt sank to Earth's center to become its core, and the silicate melt rose upward to become Earth's crust and mantle. In other words, Earth unmixed itself. The separation of silicate minerals and metals into a rocky outer layer and a metallic core, respectively, is called **differentiation**. Gravity has since pulled Earth into an almost spherical shape with a radius of 6371 km, and a circumference of about 40,000 km. However, it is not a perfect sphere, as the Earth's rotation causes an equatorial bulge, so that the Earth's circumference is 21 km (0.3%) wider at the equator than it is pole to pole. Thus it is technically an "oblate spheroid."

If we were to take an inventory of the elements that make up Earth, we would find that 95% of Earth's mass comes from only four elements: oxygen, magnesium, silicon, and iron. Most of the remaining 5% comes from aluminum, calcium, nickel, hydrogen, and sulphur. We know that the Big Bang made hydrogen, helium, and lithium, but where did the rest of the elements come from? The answer is that the other elements were made by stars. The heat and pressure within stars cause smaller atoms to smash together and fuse into new, larger atoms. For example, when hydrogen atoms smash together and fuse, helium is formed. Large amounts of energy are released when some atoms fuse and that energy is what causes stars to shine.

It takes larger stars to make elements as heavy as iron and nickel. Our Sun is an average star; after it uses up its hydrogen fuel to make helium, and then some of that helium is fused to make small amounts of beryllium, carbon, nitrogen, oxygen, and fluorine, it will be at the end of its life. It will stop making atoms and will cool down and bloat until its middle reaches the orbit of Mars. In contrast, large stars end their lives in spectacular fashion, exploding as supernovae and casting off newly formed atoms —including the elements heavier than iron — into space. It took many generations of stars creating heavier elements and casting them into space before heavier elements were abundant enough to form planets like Earth.

HOW WERE EARTH'S ATMOSPHERE AND OCEANS FORMED

Origin of Atmosphere

- Earth is believed to have formed about 5 billion years ago.
- In the first 500 million years, a dense atmosphere emerged from the vapor and gases that were expelled during degassing of the planet's interior.
- During the cooling of the earth, gases and water vapour were released from the interior solid earth. This started the evolution of the present atmosphere. This process is called degassing.
- These gases may have consisted of hydrogen (H_2), water vapor, methane (CH_4), and carbon oxides.
- Prior to 3.5 billion years ago the atmosphere probably consisted of carbon dioxide (CO_2), carbon monoxide (CO), water (H_2O), nitrogen (N_2), and hydrogen.
- The hydrosphere was formed 4 billion years ago from the condensation of water vapor, resulting in oceans of water in which sedimentation occurred.
- The most important feature of the ancient environment was the absence of free oxygen. Evidence of such an anaerobic reducing atmosphere is hidden in early rock formations that contain many elements, such as iron and uranium, in their reduced states. Elements in this state are not found in the rocks of mid-Precambrian and younger ages, less than 3 billion years old.
- One billion years ago, early aquatic organisms called blue-green algae began using energy from the Sun to split molecules of H_2O and CO_2 and recombine them into organic compounds and molecular oxygen (O_2). This solar energy conversion process is known as photosynthesis. Some of the photosynthetically created oxygen combined with organic carbon to recreate CO_2 molecules.
- The remaining oxygen accumulated in the atmosphere, touching off a massive ecological disaster with respect to early existing anaerobic organisms. As oxygen in the atmosphere increased, CO_2 decreased.
- High in the atmosphere, some oxygen (O_2) molecules absorbed energy from the Sun's ultraviolet (UV) rays and split to form single oxygen atoms. These atoms combining with remaining oxygen (O_2) to form ozone (O_3) molecules, which are very effective at absorbing UV rays. The thin layer of ozone that surrounds Earth acts as a shield, protecting the planet from irradiation by UV light.
- The amount of ozone required to shield Earth from biologically lethal UV radiation, wavelengths from 200 to 300 nanometers (nm), is believed to have

been in existence 600 million years ago. At this time, the oxygen level was approximately 10% of its present atmospheric concentration.

- Prior to this period, life was restricted to the ocean. The presence of ozone enabled organisms to develop and live on the land. Ozone played a significant role in the evolution of life on Earth and allows life as we presently know it to exist.

Three Phases of Atmosphere Formation

- *Just formed Earth*: Like Earth, the hydrogen (H_2) and helium (He) was very warm. These molecules of gas moved so fast they escaped Earth's gravity and eventually all drifted off into space.
- Earth's original atmosphere was probably just hydrogen and helium because these were the main gases in the dusty, gassy disk around the Sun from which the planets formed. The Earth and its atmosphere were very hot. Molecules of hydrogen and helium move really fast, especially when warm. Actually, they moved so fast they eventually all escaped Earth's gravity and drifted off into space.

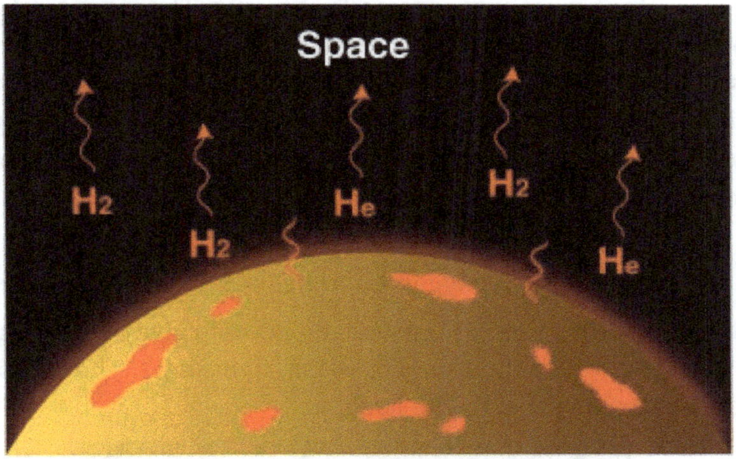

Fig. 10. Atmosphere #1

- *Young Earth*: Volcanoes released gases H_2O (water) as steam, carbon dioxide (CO_2), and ammonia (NH_3). Carbon dioxide dissolved in seawater. Simple bacteria thrived on sunlight and CO_2. By-product is oxygen (O_2).
- Earth's "second atmosphere" came from Earth itself. There were lots of volcanoes, many more than today because Earth's crust was still forming. The volcanoes released
- steam (H_2O, with two hydrogen atoms and one oxygen atom),

- carbon dioxide (CO_2, with one carbon atoms and two oxygen atoms),
- ammonia (NH_3, with one nitrogen atom and tree hydrogen atoms).

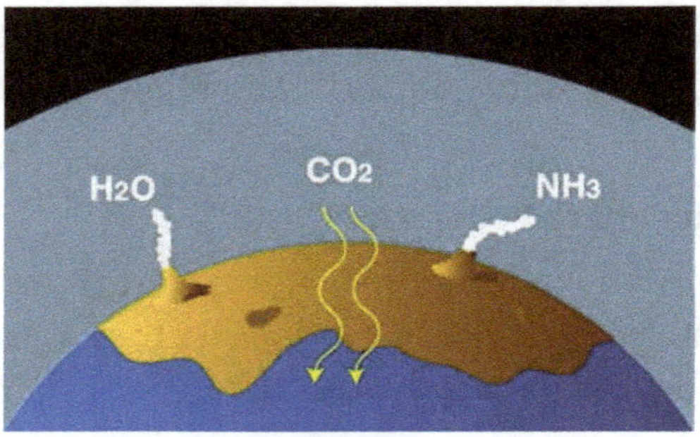

Fig. 11. Atmosphere #2

- *Current Earth:* Plants and animals thrive in balance. Plants take in carbon dioxide (CO_2) and give off oxygen (O_2). Animals take in oxygen (O_2) and give off CO_2. Burning stuff also gives off CO_2.
- Much of the CO_2 dissolved into the oceans. Eventually, a simple form of bacteria developed that could live on energy from the Sun and carbon dioxide in the water, producing oxygen as a waste product. Thus, oxygen began to build up in the atmosphere, while the carbon dioxide levels continued to drop. Meanwhile, the ammonia molecules in the atmosphere were broken apart by sunlight, leaving nitrogen and hydrogen. The hydrogen, being the lightest element, rose to the top of the atmosphere and much of it eventually drifted off into space.

Now we have Earth's "third atmosphere," the one we all know and love—an atmosphere containing enough oxygen for animals, including ourselves, to evolve.

So, Plants and some bacteria use carbon dioxide and give off oxygen, and animals use oxygen and give off carbon-dioxide—*how convenient!* The atmosphere upon which life depends was created by life itself.

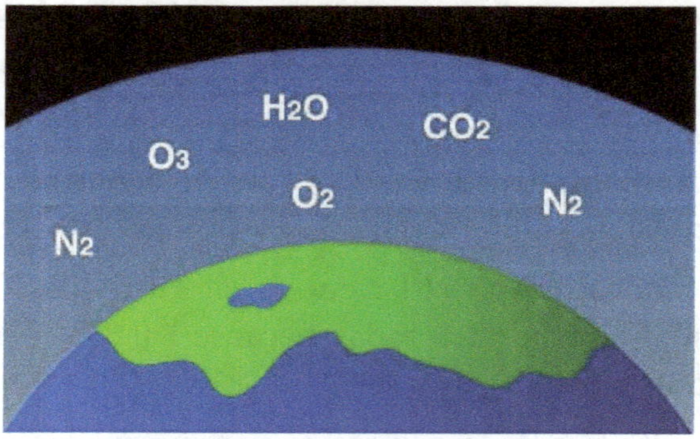

Fig. 12. Atmosphere #3

ORIGIN OF THE OCEANS

The materials that accreted on the early Earth contained the components that would eventually become our oceans and atmosphere. Under the high pressures found in the Earth's interior, gases remain dissolved in magma. As these magmas rise to the surface through volcanic activity, the pressure is reduced and the gases are released through a process called outgassing. Volcanic activity releases many different gases, including water vapor, carbon dioxide (CO_2), sulfur dioxide (SO_2), carbon monoxide (CO), hydrogen sulfide (H_2S), hydrogen gas, nitrogen, and methane (CH_4). Lighter gases such as hydrogen and helium dissipated into space, but the heavier gases remained and formed Earth's early atmosphere.

As the early Earth cooled, the water vapor in the atmosphere condensed and fell as rain. By about 4 billion years ago, the first permanent accumulations of water were present on Earth, forming the oceans and other bodies of water. Water moves between these different reservoirs through the hydrological cycle. Water is evaporated from the oceans, lakes, streams, the surface of the land, and plants (transpiration) by solar energy (Fig. 13).

It is moved through the atmosphere by winds and condenses to form clouds of water droplets or ice crystals. It comes back down as rain or snow and then flows through streams and rivers, into lakes, and eventually back to the oceans. Water on the surface and in streams and lakes infiltrates the ground to become groundwater. Groundwater slowly moves through the rock and surface materials; some returns to other streams and lakes, and some goes directly back to the oceans.

Water is stored in various reservoirs as it moves through this cycle. The largest, by far, is the oceans, accounting for 97% of the volume (Fig. 14). Of course, that water is salty. The remaining 3% is fresh water. Two-thirds of our fresh water is stored in the ground and one-third is stored in ice. The remaining fresh water — about 0.03% of the total — is stored in lakes, streams, vegetation, and the atmosphere.

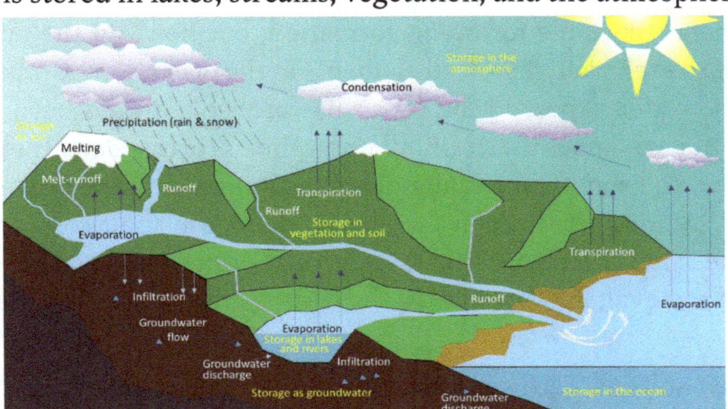

Fig. 13. Earth's hydrological cycle

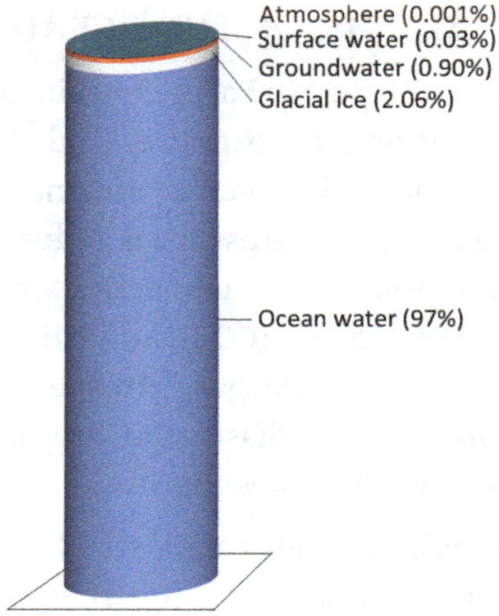

Fig. 14. Proportions of Earth's water found in the various reservoirs

To put that in perspective, let's think about putting all of Earth's water into a 1 L jug. We start by almost filling the jug with 970 ml of water and 34 g of salt. Then we add one regular-sized (~20 mL) ice cube (representing glacial ice) and two teaspoons (~10 mL) of groundwater. All of the water that we see around us in lakes and streams and up in the sky can be represented by adding three more drops from an eyedropper.

Although the proportion of Earth's water that is in the atmosphere is tiny, the actual volume is huge. At any given time, there is the equivalent of approximately 13,000 km^3 of water in the air in the form of water vapor and water droplets in clouds. Water is evaporated from the oceans, vegetation, and lakes at a rate of 1,580 km^3 per day, and just about exactly the same volume falls as rain and snow every day, over both the oceans and land. The precipitation that falls on land goes back to the ocean in the form of stream flow (117 km^3/day) and groundwater flow (6 km^3/day).

How did the oceans get salty?

Outgassing was responsible for ocean formation, but how did the ocean water get salty? Most of the salts and dissolved elements in the ocean were probably outgassed along with the water vapor, so the ocean has probably always been about as salty as it is now. But we know that rainfall and other processes weather rocks on the Earth's surface, and runoff carries dissolved substances into the ocean, contributing to its salinity. Yet despite this constant input, the ocean's salt composition remains essentially the same. Therefore, the rate of input of new material must be balanced by the rate of removal; in other words, the oceans are in a steady state in regards to salinity.

There are multiple pathways through which dissolved ions enter the ocean; runoff from streams and rivers, volcanic activity, hydrothermal vents, dissolution or decay of substances in the ocean, and groundwater input. Ions are removed from seawater as they are incorporated by living organisms (for example in shell production) or sediments, sea spray, percolation of water into the crust, or when sea water gets isolated from the ocean and evaporates.

The relationship between the input and removal of an ion can be examined through the concept of residence time, which is the average length of time a single atom of an element remains in the ocean before being removed. Residence time is calculated as:

There is great variation in residence times for different substances (Table 1). Generally speaking, substances that are readily used in biological processes have short residence times, as they are used up as they become available. Substances with longer residence times are less reactive, and may be a part of long-scale geological cycles.

Table 1 Residence times for some constituents of sea water

Constituent	Residence time (years)
Chloride (Cl-)	100,000,000
Sodium (Na+)	68,000,000
Calcium (Ca2+)	1,000,000
Water	4100
Iron (Fe)	200

So what about lakes? They are subjected to runoff and river input, so why aren't they salty like the oceans? One reason is that compared to the oceans, lakes and ponds are relatively temporary phenomena, so they do not last long enough to accumulate the same levels of ions as the oceans. Furthermore, lakes often have rivers flowing both into and out of them, so many ions are removed through the outflow, eventually finding their way to the oceans. The oceans only receive river input; there are no rivers flowing out of the ocean to remove these materials, so they are found in greater abundance in sea water. It should be noted that there are some lakes that contain water whose salt content may rival or exceed that of the ocean; these lakes usually lack river outflow. The Great Salt Lake in the western United States is an example.

CHAPTER

PLATE TECTONICS AND THE OCEAN FLOOR 3

INTRODUCTION TO PLATE TECTONICS

The surface of the Earth is divided into rigid blocks, called **tectonic plates** that are bounded by narrow regions of high deformation called plate boundaries. Tectonic plates are comprised of both crust (oceanic or continental) and mantle rock and owe their rigidity to the stiffness of mantle rock at low temperatures. Tectonic plates come in various shapes and sizes and are continuously changing shape, either through the addition of new crust and lithosphere at mid-ocean spreading centers, or through the loss of material at subduction zones. Tectonic plate shape can also change due to breaking of the plate or accretion of a piece of an adjacent plate.

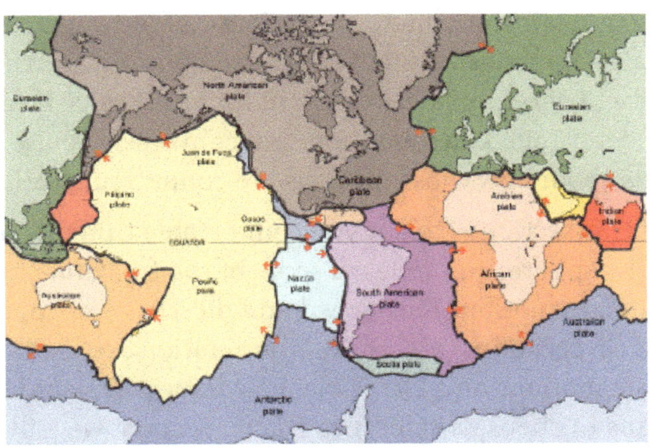

Fig. 1: Plates of the World

Tectonic plates move at rates of a few to tens of centimeters per year. However, just as the shaped of the plates are continuously changing, so to are the motion of the plates - both in terms of speed and direction. The complex shapes of the tectonic plates, the geometric constraints of plates moving on the confined surface of a sphere and changes in the forces acting on the plates all lead to changes in plate motion over time. The motion of tectonic plates are primarily driven by gravitational buoyancy forces associated with sinking of cold material into the mantle at subduction zones and rising of hot material at mid-ocean spreading centers. This motion is resisted by viscous forces on the base of the plates, by the strength of the plate itself, which resists bending into the mantle at subduction zones, and by the frictional and viscous forces acting between adjacent plates.

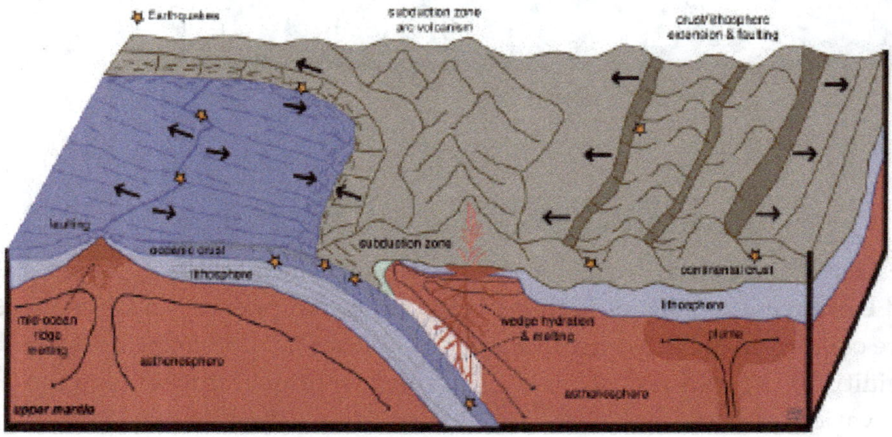

Fig. 2: Plate Tectonics

Detailed observation of both the present and past motions of tectonic plates are essential for addressing many questions in the geosciences, such as understanding the forces driving plate tectonics, the origins of intra-plate deformation, how deformation at plate boundaries manifests as earthquakes, and the physical structure of the deep mantle. Therefore, it is important for geophysicists to be able to both use observations to determine plate motions, and conversely to use to plate motions to make predictions about how those motions are connected to other observations.

The goals of this section is to get you comfortable thinking about plate motions on the surface of a sphere, and to learn some of the fundamental quantitative tools for determining plate motions (i.e., speed and direction) or to use plate motions to calculate related observations (e.g., the relative motion between two points). First, we will consider plate motions on a 2-D plane, how they are related to plate boundaries, and what it means to choose a reference frame. Second, we will review some of the ways in which we can observe plate motions, such as using magnetic anomalies on the seafloor or GPS measurements. Third, we will move to defining plate motions on a sphere using Euler poles and how to determine an Euler pole from observations. Finally, we will learn how to use a model of present-day plate motion to answer questions related to relative plate motions (e.g., How long until San Diego is at the same latitude as Sacramento?) and changes in plate boundaries (e.g., How long until the Juan de Fuca ridge reaches the Cascadia subduction zone?).

THE OCEAN FLOOR

Geography is an interesting subject which covers land (lithosphere), water (hydrosphere) and air (atmosphere). Hope you have gone through our easy-to-learn notes covering important land features. Now let's see how to learn fast the concepts associated with hydrosphere. Let's start from ocean floor first.

Earth – the water planet

- We know that our earth is the only planet in the solar system which has water in abundance; hence, it is often called as 'water planet'.
- About 71% of the earth's surface is covered by water ($3/5^{th}$ of Northern Hemisphere and $4/5^{th}$ of Southern Hemisphere is covered by water).
- Similar to the continental landforms which have been discussed in previous articles, oceans also contain a variety of landforms.

Oceanic land-forms (submarine relief)

- There are mountains, basins, plateaus, ridges, canyons and trenches beneath the ocean water too.
- These relief features found on the ocean floor are called **Submarine Relief**.

The ocean basins are broadly divided into four major subdivisions. They are:

1. Continental Shelf
2. Continental slope
3. Abyssal plains
4. The ocean deeps/ submarine trenches

Continental Shelf

- There is no clear or well-defined line separating oceans from the continent. In fact, continents do not end abruptly at the shoreline.
- They slope seaward from coast to a point where the slope becomes very steep.
- This shallow submerged extension of the continent is called as the Continental Shelf.
- The depth usually ranges from 120 to 370 meters from the surface water.
- Their width also varies from a few kilometers to more than 100 kilometers.

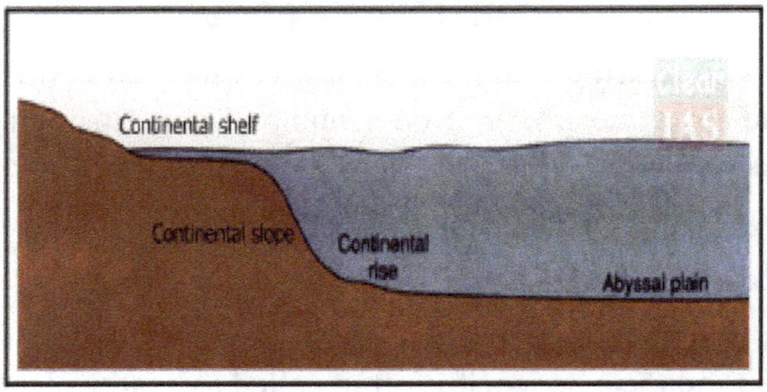

Fig. 3. Continental Shelf

- This variation can be seen even in the context of the Indian peninsula. The continental shelf of the western coast of India is much wider than that of the eastern coast.
- Continental shelves are much narrow or absent in some continents, particularly where Fold Mountains run parallel or close to the coast.
- The shallow water over the shelf enables sunlight to penetrate through the water to the bottom and encourages the growth of microscopic plants and animals called planktons, which are the food for fishes.
- Continental Shelves are of greater importance to man. They are the sources of fishes, minerals including sand and gravel, etc.
- A large quantity of world's petroleum and natural gas are obtained from the shelves.
- Coral reefs are also common on continental shelves.
- One of the striking features of the continental shelf is the presence of submarine canyons which extend to the continental slope.
- These canyons are 'steep-sided valleys' cut into the floor of the seas.
- They are very similar to the gorges found on the continents.
- One of the reasons for the formation of the submarine canyon is the underwater landslide.
- The continental shelf is generally considered to be territorial water extents of the nations to which it adjoins.

Continental Slope

- The continuous sloping portion of the continental margin, seaward of the continental shelf and extending down to the deep sea floor of the abyssal plain, is known as Continental Slope.
- It extends between the depths of 180m to 3600m.

- They have very little deposits of sediments on them due to their steepness and increasing distance from the land.
- Sea life is also far less here than on the shelf.
- The base of the continental slope will have deposits of sediments. This belt of sedimentary deposits forms the **Continental Rise.**

Abyssal Plain

- Abyssal plains are the extremely flat and featureless plains of the deep ocean floor.
- Abyssal plains are considered as the levelest areas on the earth.
- They cover a major portion of the ocean floors between the depths of 3000m to 6000m.
- Abyssal plains are as irregular as the continental plains with submarine plateaus, hills, guyots, and seamounts.
- The floor of the abyssal plain is covered by sediments.
- The sediments which are formed from the remains of living things are called **Oozes.**
- Oozes can be seen in those seas which favour an abundant growth of organisms.
- Another type of sediments is **red clay** which is of volcanic origin or made up of tiny particles brought by wind and rivers.
- Some relief feature seen on abyssal plains are:

Submarine Ridges

- The oceanic mountains are called as submarine ridges.
- They are linear belts occurring near the middle of the oceans and are also called mid-oceanic ridges.
- The oceanic ridge is the site of frequent earthquakes.
- Volcanism is common in ocean ridges and it produces many relief features.
- The Mid-Atlantic Ridge is the largest continuous submerged mountain ridge which runs from north to south in the Atlantic Ocean.
- At some places, the peaks rise above the sea surface to form Islands.

Seamounts and Guyots

- Submerged volcanoes with sharp tops are called as seamounts.
- Sometimes they rise above the sea as isolated islands.

- Hawaii and Tahiti Islands are such exposed tops of volcanoes.
- Seamounts with a flattened top are called as Guyots or Tablemounts.

The Ocean Deeps/ Submarine Trenches

- The ocean deeps are the deepest part of the ocean.
- They are generally called submarine trenches.
- These trenches are not always located in the middle of the ocean basins. They may be very close or parallel to the continents bordered by Fold Mountains.
- They are usually found adjacent to the areas of volcanic and earthquake activity.
- Great earthquakes and tsunamis are born in them.
- The Pacific Ocean has the largest number of trenches.
- The Mariana trench in the Pacific Ocean is the deepest known part of the world.

Bathymetry, the shape of the ocean floor, is largely a result of a process called plate tectonics. The outer rocky layer of the Earth includes about a dozen large sections called tectonic plates that are arranged like a spherical jig-saw puzzle floating on top of the Earth's hot flowing mantle. Convection currents in the molten mantle cause the plates to slowly move about the Earth a few centimeters each year. Many ocean floor features are a result of the interactions that occur at the edges of these plates.

The shifting plates may collide (converge), move away (diverge) or slide past (transform) each other. As plates converge, one plate may move under the other causing earthquakes, forming volcanoes, or creating deep ocean trenches. Where plates diverge from each other, molten magma flows upward between the plates, forming mid-ocean ridges, underwater volcanoes, hydrothermal vents, and new ocean floor crust. Transform boundaries are faults that connect two areas where plates are converging or diverging. The edges of these continental boundaries usually form zig-zag patterns.

Hydrography

Hydrography is the science that measures and describes the physical features of bodies of water and the surrounding land. In 1807, President Thomas Jefferson established the U.S. Coast Survey, the predecessor to NOAA, to conduct hydrographic surveys and create nautical charts of the young nation's ports and waterways.

Historically, ships would take depth "soundings" by dropping a lead line off of a ship until it reached the bottom and the line went slack. The term "soundings" comes from the Old French "sonder"offsite link (to plumb) or Old English "sund"offsite link for swimming, water, or sea. But coincidentally, this work is now usually done using sound! Today, ships equipped with multibeam echo sounders are the primary method of gathering this information, although hydrographers use many other technologies as well.

By mapping water depth, the shape of the seafloor and coastline, and the location of possible obstructions, hydrography helps to keep our maritime transportation system moving safely and efficiently, both at sea and in the Great Lakes. Hydrography is also used to identify shipwrecks and preserve our maritime heritage.

Very little of the ocean floor has been mapped directly. So how do we make maps of the global ocean floor? It turns out that satellites can "see" below the sea surface. With careful processing, small differences in sea surface heights and gravity can reveal detailed maps of the seafloor.

CONTINENTAL DRIFT THEORY AND EVIDENCES

The Matching of Continents (Jig-Saw-Fit)

- The coastlines of South America and Africa fronting each other have a remarkable and unique match.
- In 1964, Bullard created a map using a computer program to find the right fit of the Atlantic margin and it proved to be quiet.
- Rocks of the Same Age across the Oceans
- The radiometric dating methods have helped in correlating the formation of rocks present in different continents across the ocean.
- The ancient rocks belts on the coast of Brazil match with those found in Western Africa.
- The old marine deposits found in the coasts of South America and Africa belong to the Jurassic Age. This implies that the ocean never existed before that time.

Tillite

- It is the sedimentary rock made from glacier deposits.
- The Gondwana system of sediments from India is recognized as having its counterparts in 6 different landmasses in the Southern Hemisphere.
- Counterparts of this series are found in Madagascar, Africa, Antarctica, Falkland Island, and Australia not to mention India.
- At the base, the system has thick tillite signifying widespread and sustained glaciation.
- Generally, the similarity of the Gondwana type sediments shows that these landmasses had exceptionally similar origins.

- The glacial tillite gives clear evidence for palaeoclimates and the drifting of continents.

Placer Deposits

- The presence of abundant placer deposits of gold along the Ghana coast and the complete lack of its source rocks in the area is a phenomenal fact.
- The gold-bearing veins are present in Brazil and it is evident that the gold deposits of Ghana in Africa are obtained from the Brazil plateau from the time when the two continents were beside each other.
- The widespread distribution of Permo-Carboniferous glacial sediments in South America, Africa, Madagascar, Arabia, India, Antarctica, and Australia was one of the major pieces of evidence for the theory of continental drift.
- The continuity of glaciers, inferred from oriented glacial striations and deposits called tillites, suggested the existence of the supercontinent of Gondwana, which became a central element of the concept of continental drift.

Distribution of Fossils

- The interpretations that Lemurs occur in India, Africa, and Madagascar led to the theory of a landmass named "Lemuria" connecting these 3 landmasses.
- Mesosaurus was a tiny reptile adapted to shallow brackish water.
- The skeletons of these creatures are found in the Traver formations of Brazil and Southern Cape Province of South Africa.

Evidence for continental drift

Besides the way the continents fit together, Wegener and his supporters collected a great deal of evidence for the continental drift hypothesis.

- Identical rocks, of the same type and age, are found on both sides of the Atlantic Ocean. Wegener said the rocks had formed side-by-side and that the land had since moved apart.
- Mountain ranges with the same rock types, structures, and ages are now on opposite sides of the Atlantic Ocean. The Appalachians of the eastern United States and Canada, for example, are just like mountain ranges in eastern Greenland, Ireland, Great Britain, and Norway. Wegener concluded that they formed as a single mountain range that was separated as the continents drifted.
- Ancient fossils of the same species of extinct plants and animals are found in rocks of the same age but are on continents that are now widely separated.

Wegener proposed that the organisms had lived side by side, but that the lands had moved apart after they were dead and fossilized. He suggested that the organisms would not have been able to travel across the oceans.

- Fossils of the seed fern *Glossopteris* were too heavy to be carried so far by wind.
- *Mesosaurus* was a swimming reptile but could only swim in fresh water.
- *Cynognathus* and *Lystrosaurus* were land reptiles and were unable to swim

- Grooves and rock deposits left by ancient glaciers are found today on different continents very close to the equator. This would indicate that the glaciers either formed in the middle of the ocean and/or covered most of the Earth. Today glaciers only form on land and nearer the poles. Wegener thought that the glaciers were centered over the southern land mass close to the South Pole and the continents moved to their present positions later on.
- Coral reefs and coal-forming swamps are found in tropical and subtropical environments, but ancient coal seams and coral reefs are found in locations where it is much too cold today. Wegener suggested that these creatures were alive in warm climate zones and that the fossils and coal later had drifted to new locations on the continents.

EVIDENCE OF CONTINENTAL DRIFT THEORY

Fossil plants and animals clue

Fossils of creatures and plants discovered on different continents helped to him push his case that the continents were once locked together before breaking up and drifting away. Some of the fossil evidence discovered on the continents includes mesosaurus, lystosaurus, cygnognathus and glossopteris.

Mesosaurus

Mesosaurus are fossils originating from aquatic freshwater reptiles. These reptiles lived in freshwater bodies such as lakes and rivers. If you take a look at the map before the continents drifted away, you find that Mesosaurus is located in the southern parts of Africa and North America. If you look at the current map after the continents drifted away, you find that North America and Africa are far away from each other separated by the Atlantic Ocean.

Lystrosaurus and Cygnognathus

Lystrosaurus is Fossil from land reptiles. Cygnognathus is Fossil from another species of reptile. If you look at the map before the continents drifted away, you find that Cygnognathus was situated in South America and Africa. Lystrosaurus, on the other hand, was located in Africa, India, and Antarctica. According to the Lystrosaurus and Cygnognathus' body anatomy, they are not designed for swimming. Now, looking at the current map after drifting of the continents, Cygnognathus is situated in Africa and South America.

So, technically, the creature would not have traveled through the ocean to get to the next continent. The distribution of these reptiles in different continents proves that these separate land masses were once together before splitting. On top of that, Lystrosaurus is situated in Antarctica, India, and Africa according to the modern map. Again, for this creature to swim from Africa to Antarctica would have been impossible. So, the fact that these organisms could not swim through the vast water bodies helps to substantiate the fact that the continents were once together to allow distribution of these organisms in various continents where their fossils were found on.

Glossopteris

Glossopteris was Fossil from fern. It is a fern that was discovered in Africa, South America, Antarctica, India, and Australia. According to the modern day map, these continents are far apart. It's, therefore, impossible that the seeds of the fern were blown by wind or floated across the ocean to have that kind of population on these 5 continents. The fact that the seeds would have been unable to be blown or floated to these continents proves that these continents were once together.

Information from climate

Apart from fossil evidence, Wegener used climatic clues to substantiate his facts. One of the clues he used was the glacier. Wegener, in his expedition, discovered glacier grooves in all the continents. Glacier grooves are the gaping trenches or landmarks carved out by movement of glacier. He discovered glacier groves in continents that were not cold enough to support snow formation in the present day. When he brought these continents together, the glacier grooves perfectly matched up. This suggested that these continents were once together and drifted away over time.

The perfect fitting of continents

Wegener also provided evidence of continental drift by looking at the physical shape of continental coastlines. The shape of continental coastlines visibly matched up like pieces of the puzzle. For example, the continents of South America and Africa fit perfectly when brought together. This indicates that the continents were once fixed together and drifted apart over time.

Rock clues

Wegener specifically used rock composition and mountain ranges to provide evidence of continental drift. He provided evidence that mountain ranges occurring on opposite continents matched up pretty well when brought together. This perfect aligning of mountain ranges indicated that the continents were once together. For example, if you take the mountains ranges of North America and Europe, they align perfectly well. Mountains are typically formed when two opposite land masses collide with each other triggering the rock found at the point of collision to fold or bulge upwards.

In terms of rock composition, at locations where continents match up, the kinds of rocks on the opposite sides of the continents were of the same kind. Since these match up points consisted of the same kind of rock further proves that the continents were once together before breaking away and drifting.

Despite Wegener's effort to try to prove his theory of continental drift, it was never accepted. When he was tasked with proving how these continents moved to different places, he couldn't. However, this theory of continental drift came to be accepted in the 90's after his death. He died of an alleged heart attacked when on an expedition to Greenland.

Magnetic polarity evidence

Puzzling new evidence came in the 1950s from studies on the Earth's magnetic history (Fig. 4). Scientists used **magnetometers**, devices capable of measuring the magnetic field intensity, to look at the magnetic properties of rocks in many locations.

Magnetite crystals are like tiny magnets that point to the north magnetic pole as they crystallize from magma. The crystals record both the direction and strength of the **magnetic field** at the time. The direction is known as the field's **magnetic polarity.**

Magnetic Polarity on the Same Continent with Rocks of Different Ages

Geologists noted important things about the magnetic polarity of different aged rocks on the same continent:

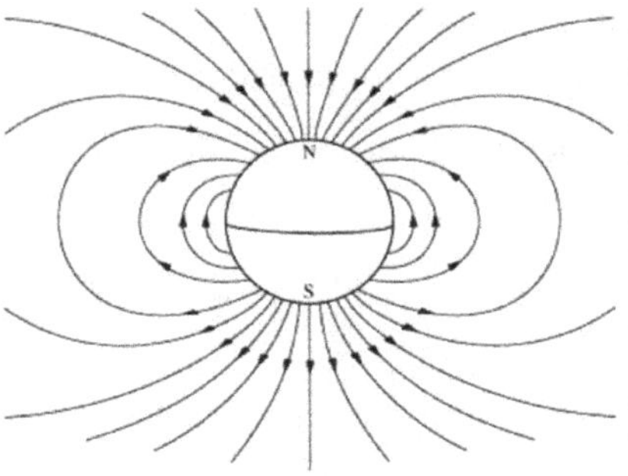

Fig. 4. Earth's magnetic field is like a magnet with its north pole near the geographic North Pole and the south pole near the geographic South Pole.

- Magnetite crystals in fresh volcanic rocks point to the current magnetic north pole no matter what continent or where on the continent the rocks are located.
- Older rocks that are the same age and are located on the same continent point to the same location, but that location is not the current north magnetic pole.
- Older rock that are of different ages do not point to the same locations or to the current magnetic north pole.

In other words, although the magnetite crystals were pointing to the magnetic north pole, the location of the pole seemed to wander. Scientists were amazed to find that the north magnetic pole changed location through time (Fig. 5).

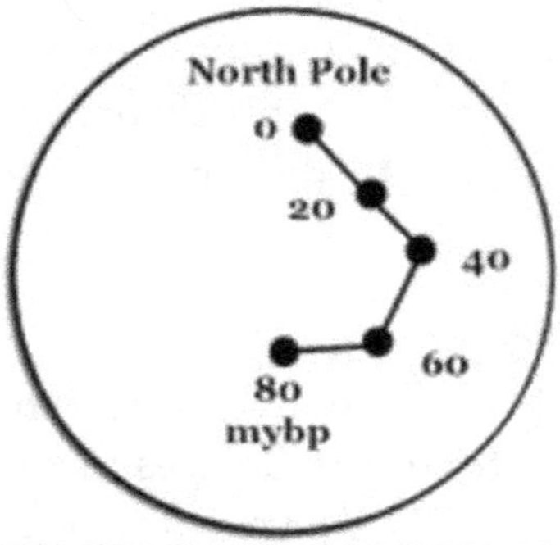

Fig. 5. Earth's Apparent Polar Wander: The location of the north magnetic north pole 80 million years before present (mybp), then 60, 40, 20, and now.

There are three possible explanations for this:

1. The continents remained fixed and the north magnetic pole moved.
2. The north magnetic pole stood still and the continents moved.
3. Both the continents and the north pole moved.

Magnetic Polarity on Different Continents with Rocks of the Same Age

Geologists noted that for rocks of the same age but on different continents, the little magnets pointed to different magnetic north poles.

- 400-million-year-old magnetite in Europe pointed to a different north magnetic pole than the same-aged magnetite in North America.
- 250 million years ago, the north poles were also different for the two continents.

The scientists looked again at the three possible explanations. Only one can be correct. If the continents had remained fixed while the north magnetic pole moved, there must have been two separate north poles. Since there is only one north pole today, the only reasonable explanation is that the north magnetic pole has remained fixed but that the continents have moved.

To test this, geologists fitted the continents together as Wegener had done. It worked! There has only been one magnetic north pole and the continents have drifted. They named the phenomenon of the magnetic pole that seemed to move but actually did not apparent polar wander.

Historical Observations leading up to Plate Tectonics Theory

Much of the background work related to the modern theory is the culmination of hundreds of years of world-wide exploration and geologic observations that go back several centuries. Highlights include:

Early Maps of the World: Maps compiled by early global explorations resulted in the observation of the matching shapes of the coastlines on opposite sides of the Atlantic Ocean. These similarities were noted from early maps by by a Flemish cartographer named **Abraham Ortelius** in 1596 who first suggested that it looked like the continents had drifted apart. Geographers preparing maps noted that charts of land masses that the shape continents of the world fit together like a jig-saw puzzle. It took centuries before an organized effort started to analyze the data!

Observations of the location of the **world's volcanoes** (maps) and lead to the recognition of the **Ring of Fire.** As we will see in the next chapter, the volcanoes of the Ring of Fire relate directly to Plate Tectonics Theory.

Continental Drift

The Continental Drift Hypothesis was a prelude to the modern Plate Tectonics Theory.

The Continental Drift Hypothesis was proposed by a German astronomer and meteorologist named Alfred Wegener (1880-1930), but based on research by other earlier observers. The Continental Drift Hypothesis was based on observations that the continental coastlines on either side of the Atlantic Ocean seemed to match up. More importantly, the drift hypothesis was supported by similarities of fossils, rocks, and mountain belts on both sides of the Atlantic, as explained further below.

Wegener's hypothesis was that all the observable continents bordering the Atlantic and Indian Oceans had once assembled into a single supercontinent that he named **Pangaea.** He determined that this great landmass began to break apart about 200 million years ago. He said that these continents started moving about 200 million years ago (MYA) and they are still moving. This is amazing because Wegener proposed this more than 100 years ago and it remains correct today. However, the theory was rejected by most scientists at the time.

Pangaea—a supercontinent comprising all the continental crust of the earth, theorized to have assembled from other continental land masses in middle to late Paleozoic time. The assembled landmass, Pangaea, existed through late Paleozoic and through early Mesozoic times before the continents separated and gradually migrated into their current configuration.

Geologic and fossil evidence indicated that some land masses which are now separated may have been together at some point in time. However, nobody could provide a mechanism for the continents to move.

Fossil evidence supporting Continental Drift includes:

- Similar fossil sites occur on different continents, when Pangaea was together. Examples include Mesosaurus and Glossopteris fossils. **Mesosaurus** is a reptile from the Early Permian of southern Africa and South America. **Glossopteris** are fossilized woody plants from the Permian and Triassic Periods.
- Fossils and organic deposits in sedimentary rocks provide information about the geologic features, the climate, and flora and fauna for an area through geologic time.
- In many regions, fossils are found presently in climates and locations that could not have supported them. Example: fossil corals found in marine sediments of northern Alaska. This is because continents drifted from one climate zone to another, carrying fossils with them.
- Fossils show that marked **divergent evolution** occurred at times when the continents were split apart. This can account for the **present-day distribution** of plants and animals around the globe (especially marsupial mammals).

Geologic features supporting Continental Drift include:

- **Old mountain ranges** cross continental boundaries that are now separate but were once together. (Examples: The **Appalachian Mountains** in eastern US and **Atlas Mountain** of northwest Africa).
- The formation of rocks in climates that could not support those types of rocks. Example: Permian-age rocks displaying evidence of **glaciation** in found in southern India.

The **Sedimentary Rock Record** in many regions around the world reveals the progression of changes that took place as continental landmasses split apart. For example, When Africa pulled apart from South America the sediments reveal:

- First, the land was uplifted, **rifted apart**, and a very large lake filled in the low region.

- The rift partially opened to the ocean from the north.
- Finally, it opened completely to the open ocean to the south.
- This occurred between 60 and 100 million years ago.

WHAT FEATURES FORM AT PLATE TECTONIC BOUNDARIES?

The Earth's outer crust (the lithosphere) is composed of a series of tectonic plates that move on a hot flowing mantle layer called the asthenosphere. Heat within the asthenosphere creates convection currents that cause tectonic plates to move several centimeters per year relative to each other. When two tectonic plates meet, we get a "plate boundary." There are three **major types of plate boundaries**, each associated with the formation of a variety of geologic features.

If two tectonic plates collide, they form a convergent plate boundary. Usually, one of the converging plates will move beneath the other, a process known as subduction. Deep trenches are features often formed where tectonic plates are being subducted and earthquakes are common at subduction zones as well. As the sinking plate moves deeper into the mantle, fluids are released from the rock causing the overlying mantle to partially melt. The new magma (molten rock) rises and may erupt violently to form volcanoes, often building arcs of islands along the convergent boundary.

When two plates are moving away from each other, we call this a divergent plate boundary. Along these boundaries, magma rises from deep within the Earth and erupts to form new crust on the lithosphere. Most divergent plate boundaries are underwater and form submarine mountain ranges called oceanic spreading ridges. While the process of forming these mountain ranges is volcanic, volcanoes and earthquakes along oceanic **spreading ridges** are not as violent as they are at convergent plate boundaries.

The third type of plate boundary occurs where tectonic plates slide horizontally past each other. This is known as a transform plate boundary. As the plates rub against each other, huge stresses can cause portions of the rock to break, resulting in earthquakes. Places where these breaks occur are called faults. A well-known example of a transform plate boundary is the San Andreas Fault in California.

PLATE TECTONICS AND SEA LEVEL CHANGE

Today, the Earth's ocean is made up of the large Pacific, Atlantic, Indian, and Arctic Oceans. These bodies of water were not always in their current shape and configuration. As a result, you can imagine the large-scale changes in sea level that would have accompanied their assembly since the last super-continent (Pangea) began to break up some 250 million years ago. These changes would have been very slow but significant, operating on time scales beyond those experienced by human beings.

Long-Term Sea Level Change (hundreds of thousands to millions of years) is influenced by factors that modify the size and shape of ocean basins. Global or eustatic sea level can change as the result of changes in the number, size, and shape of ocean basins. Throughout Earth's history, the global ocean has been modified by plate tectonics. Often, large continents assembled from smaller ones produced more expansive oceans between them. These expansive ocean bodies were subsequently dissected when super-continents rifted and formed smaller oceans out of the formerly vast oceans.

The tectonic processes at work on the Earth influence the size of ocean basins and therefore, sea level in many, complex ways. The following list gives an idea of some of these processes and their interactions and feedback mechanisms:

- rifting of tectonic plates at divergent plate boundaries;
- assembly of micro-continents, volcanic terrains, continents - especially supercontinents like Rodinia, Pangea, etc.;
- subduction of tectonic plates at ocean trenches at convergent plate boundaries;
- eruption and formation of large igneous provinces that originate from massive extrusions of lava, oceanic plateaus, hotspot volcanic island chains, etc.;
- high rates of volcanism on the seafloor volumetrically displace water out of the ocean basin producing higher sea levels (called transgression of sea level);
- low rates of volcanism allow water to return to the ocean basin and sea levels drop (called regression of sea level);
- when rocks cool from a molten state, they contract in volume; this allows subsidence to occur, especially along the mid-ocean ridges, and sea levels fall;
- when rates of volcanism are low, rocks tend to cool faster and sea levels drop as subsidence occurs.
- conversely, when rates of volcanism are high, it takes longer for the rocks to cool, and sea level remains higher for longer periods of time after the rate of volcanism subsides.

Different types of plate tectonic boundaries

The Earth's lithosphere, which includes the crust and upper mantle, is made up of a series of pieces, or tectonic plates, that move slowly over time.

A divergent boundary occurs when two tectonic plates move away from each other. Along these boundaries, earthquakes are common and magma (molten rock) rises from the Earth's mantle to the surface, solidifying to create new oceanic crust. The Mid-Atlantic Ridge is an example of divergent plate boundaries.

When two plates come together, it is known as a convergent boundary. The impact of the colliding plates can cause the edges of one or both plates to buckle up into a mountain ranges or one of the plates may bend down into a deep seafloor trench. A chain of volcanoes often forms parallel to convergent plate boundaries and powerful earthquakes are common along these boundaries. The **Pacific Ring of Fire** is an example of a convergent plate boundary.

At convergent plate boundaries, oceanic crust is often forced down into the mantle where it begins to melt. Magma rises into and through the other plate, solidifying into granite, the rock that makes up the continents. Thus, at convergent boundaries, continental crust is created and oceanic crust is destroyed.

Two plates sliding past each other forms a **transform plate boundary**. One of the most famous transform plate boundaries occurs at the San Andreas fault zone, which extends underwater. Natural or human-made structures that cross a transform boundary are offset — split into pieces and carried in opposite directions. Rocks that line the boundary are pulverized as the plates grind along, creating a linear fault valley or undersea canyon. Earthquakes are common along these faults. In contrast to convergent and divergent boundaries, crust is cracked and broken at transform margins, but is not created or destroyed.

Extrinsic Drivers of Sea Level Change

Earth Orbit, Solar Insolation Variability & Sea Level Change

Take look at the graphs (Fig. 6) showing changes in sea level through the history of the Earth. The top figure covers 35 million years, the middle one covers a little more than 5 million years, and the bottom one zooms in to the past 500,000 years. What do you notice? Is there a regularity in the pattern of sea level ups and downs?

Through Earth's history, there appears to have been a regular spacing of glacial maximum events, at roughly 120 thousand years (ky). When variations in Earth's orbit produce repetitive changes in climate and sea level, the observed cycles are often referred to as Milankovitch Cycles. Many sedimentary rock sequences have been shown to have stacking patterns that reflect these time scales, as do ice core data.

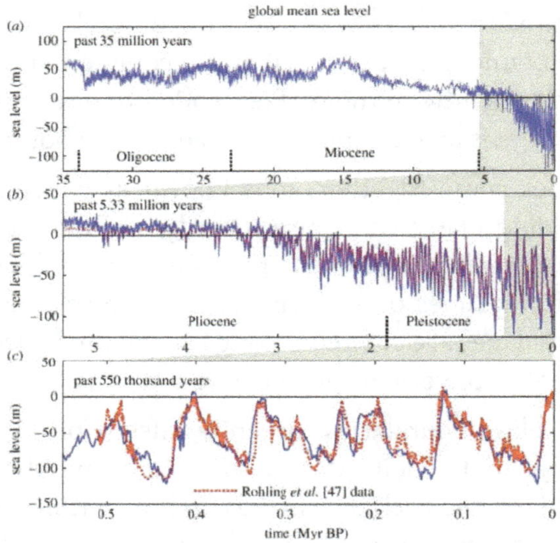

Fig. 6. Reconstruction of Sea level change over three time periods: a. 35 million years, b. 5.33 million years, c. 550,000 years.

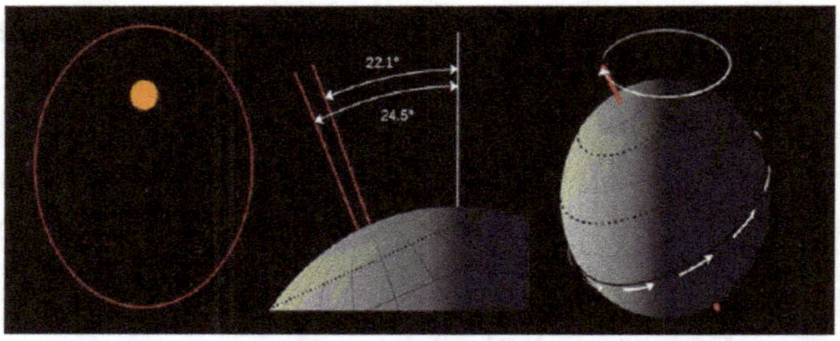

Fig. 7. This figure is a composite of diagrams designed to show each of the three types of cycles outlined here

Milankovitch Cycles

Mathematician Milutin Milankovitch proposed an explanation for the changes in the way the Earth orbits the sun. These changes define the sequence of ice ages and warm periods.

- The Earth's orbit changes from being nearly circular to slightly elliptical (eccentricity). This cycle is affected by other planets in the solar system and has a period of around 100,000 years.
- The angle of tilt of the Earth's axis changes from 22.1° to 24.5° (obliquity). This cycle has a period of 41,000 years.
- The direction of the tilt of the axis changes (precession) on a cycle of 26,000 years.
- These changes influence the length of the seasons and the amount of solar radiation received by the Earth.

CHAPTER

MARINE SEDIMENTS 4

The sediments deposited in the ocean are an archive of historical information about the Earth and, specifically, they provide information about global biogeochemical cycles. The distribution of sediments in the ocean is determined by biological and chemical processes and therefore should be discussed in the context of marine chemistry.

The first detailed study of marine sediments was done in the 1870's. An expedition called the "Challenger expedition" led by Sir Murray and Renard dredged the bottom of the ocean systematically and described the sediments. They classified their findings into five major groups: red clays, carbonate ooze, silicic ooze, nodules and volcanic material. This is quite similar to what we do today. We categorize the type of sediments mostly by their source and composition. In the late 1940's, the first recovery of a long core was achieved and that introduced the possibility of investigating sediments over time. Shortly afterward, a global ocean drilling program was designed and sediment cores are now continuously collected from different oceanic locations.

The sediments in the ocean consist of three major components: detrital, biogenic and authigenic based on their origin. They are classified using the 30% rule, if there is more than 30% of any type of component in the sediment it will be classified as such. However, it must be kept in mind that there are no pure detrital, authigenic or biogenic sediments; sediments are always mixtures of different components.

Detrital: brought into the ocean from outside, consists of terrigenous, volcanic, and cosmogenic material

Terrigenous sediments are those where the ultimate source is weathering and erosion of rocks on land. The materials composing these sediments are introduced to the ocean by water, wind or ice. Terrigenous sediments are more abundant close to the continents, specifically near river mouths and in the very deep areas of the oceans.

The composition of terrigenous sediments thus depends on the climate and weathering at the source region, and the distribution of these sediments by the geographic deposition of rivers and the prevailing winds. Rivers and glaciers carry large particles and deposit 10^{10} tons annually, while the winds deposit 10^8 tons of much smaller particles.

These sediments are most abundant on continental margins where rivers come in (Atlantic > Pacific). The smaller, wind blown terrigenous particles, are present everywhere in the oceans and they are the major component of a large fraction of the sediment in the deep Pacific basin, mostly because the biogenic components are not preserved there. Thus, terrigenous and hydrogenous minerals are the dominant components of red clays. This does not mean that the input of this material is larger there but rather that there is nothing else to dilute it with. Terrigenous sediments enclose information about river fluxes, sources of the weathered material (provenance), weathering processes on land, wind stress and direction, as well as glacial extent and glaciers location (ice rafted debris).

Volcanic sediments are composed of minerals brought into the ocean mostly by wind, as dust and ash from volcanic eruptions. They are typically in the size range of 1m. Their abundance gives information about periods and locations of intensive volcanic activity. These sediments are more abundant close to volcanic islands but also globally distributed after big eruptions. These deposits are easy to date by radioactive age determination and because of their global and instantaneous distribution can be used for global correlation.

Cosmogenic particles are those that arrive from outer space and survive the Earth's atmosphere to enter the sedimentary record. About 4-6 × 10^4 tons of these particles accumulate each year, they range in size from 0.1 to 1mm and have a typical spherule shape. They have distinct geochemical signatures (high ^3He, Ir, Os and Os isotopes and organic compounds not typical to Earth) and provide information about possible changes in the rate of cosmogenic bombardment as well as catastrophic impacts (K-T). They could also be a mean of estimating sedimentation rate if we assume constant accumulation.

Authigenic components are oceanic inorganic minerals that precipitate directly from the seawater, either in the water column or in the sediment after burial. These minerals make up only a small fraction of deep-sea sediments today, but in special environments and certain geological times, they comprise the bulk of the sedimentary sequence.

These include evaporites, which are sediments that are deposited as a result of evaporation of water and usually occur in restricted basins and lagoons on the shelf where evaporation exceeds precipitation and flushing in of water. This includes the minerals anhydrite, gypsum, halite, dolomite and more. These sediments enclose information about the climate and basin configuration conditions at the location and time of their formation and also carry valuable information about the chemical makeup of seawater. For example, when anhydrite precipitates, the sulfate forming this mineral comes from seawater and has the same isotopic composition as seawater sulfate, so we get a record of the seawater S isotopic composition.

Another group of important authigenic components are Hydrogenous Fe-Mn oxyhydroxides: These minerals usually form as coating on existing minerals, but also occur as nodules and crusts. The Fe-Mn oxides and hydroxides co-precipitate and scavenge trace metals from seawater and thus could reveal chemical information of the ocean.

Hydrothermal deposits are also Authigenic; they precipitate around active ridges where hydrothermal circulation leaches transition metals from the basalt and carries the dissolved material to the seafloor. Oxidizing conditions and cooling of these hydrothermal solutions results in precipitation of Fe and Mn oxides and hydroxides in the form of crusts or metal rich ooze, as well as sulfides ($(Zn,Fe)S$, $CuFeS_2$, FeS_2) and sulfates ($BaSO_4$, $CaSO_4$). The existence of these minerals can indicate locations of previous hydrothermal activity and carry information about the temperature and nature of the reactions with the basalt.

Biogenic Sediments are one of the most important constituents of marine sediments. As the name implies, these form directly or indirectly through biological activity. They are made of a variety of delicate and intricate structures mostly of skeletal remains of marine phytoplankton and zooplankton. The life span of most of these organisms is on the order of weeks, so there is a slow continuous "rain" of their remains down through the water column to build successive layers of sediment. The distribution of these sediments would depend on the abundance of organisms precipitating these phases and dissolution at depth (e.g. preseration).

CLASSIFYING SEDIMENTS

The term "sediment" refers to the tiny particles of rocks and other materials that sink to the ocean floor and eventually settle and accumulate on the bottom. All regions of the seafloor contain some form of sediment, although there are many different types of sediments from a variety of sources, and the amount of accumulated sediment can vary greatly from place to place. Globally, ocean sediments average about 1 km thick, but they can exceed 15 km thick in areas of high accumulation (Fig. 2). These areas include regions near the mouths of rivers where there is high sediment discharge, and passive margins near the continents where the seafloor has had millions of years for sediment to accumulate. On the other hand, sediments are sparse along divergent plate boundaries where new oceanic crust is being formed, as the crust is too new for significant accumulation, and in the central oceans that are far away from any significant sediment sources.

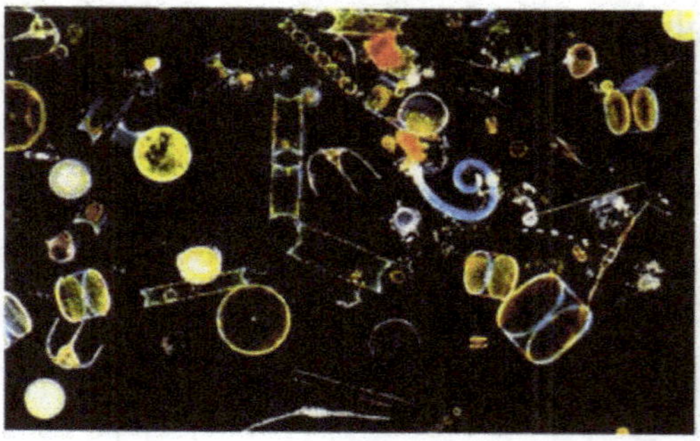

Living Phytoplankton

Living Zooplankton

Fig. 1

As time passes over millions of years, these sediments can become **lithified** or turned into sedimentary rock. It has been estimated that over half of the exposed rock on the continents is sedimentary rock originally deposited in ancient oceans and uplifted by plate tectonics. Many tall mountains, including Mt. Everest, are composed of rock formations that contain fossils of marine creatures. These rocks were originally formed as ocean sediments which were then lithified and pushed upwards during the process of mountain formation.

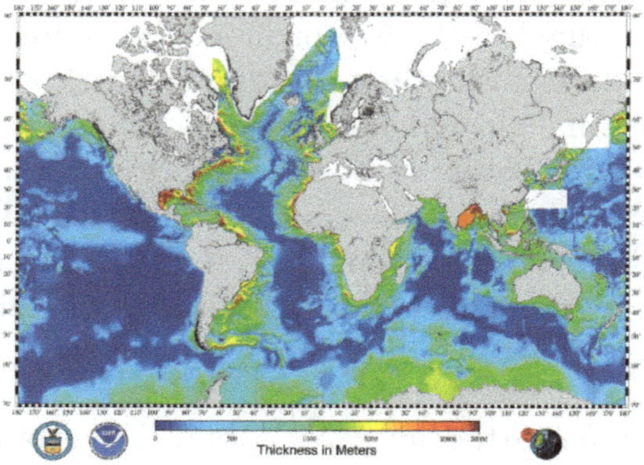

Fig. 2. Total sediment thickness of the world ocean

SEDIMENT CLASSIFICATION

There are a number of ways that we can classify ocean sediments, and some of the most common distinctions are based on the sediment texture, the sediment composition, and the sediment's origin.

Texture

Sediment texture can be examined through several variables. The first is grain size. Sediments are classified by particle size, ranging from the finest clays (diameter <0.004 mm) to the largest boulders (> 256 mm) (Fig. 3). Among other things, grain size represents the conditions under which the sediment was deposited. High energy conditions, such as strong currents or waves, usually results in the deposition of only the larger particles as the finer ones will be carried away. Lower energy conditions will allow the smaller particles to settle out and form finer sediments.

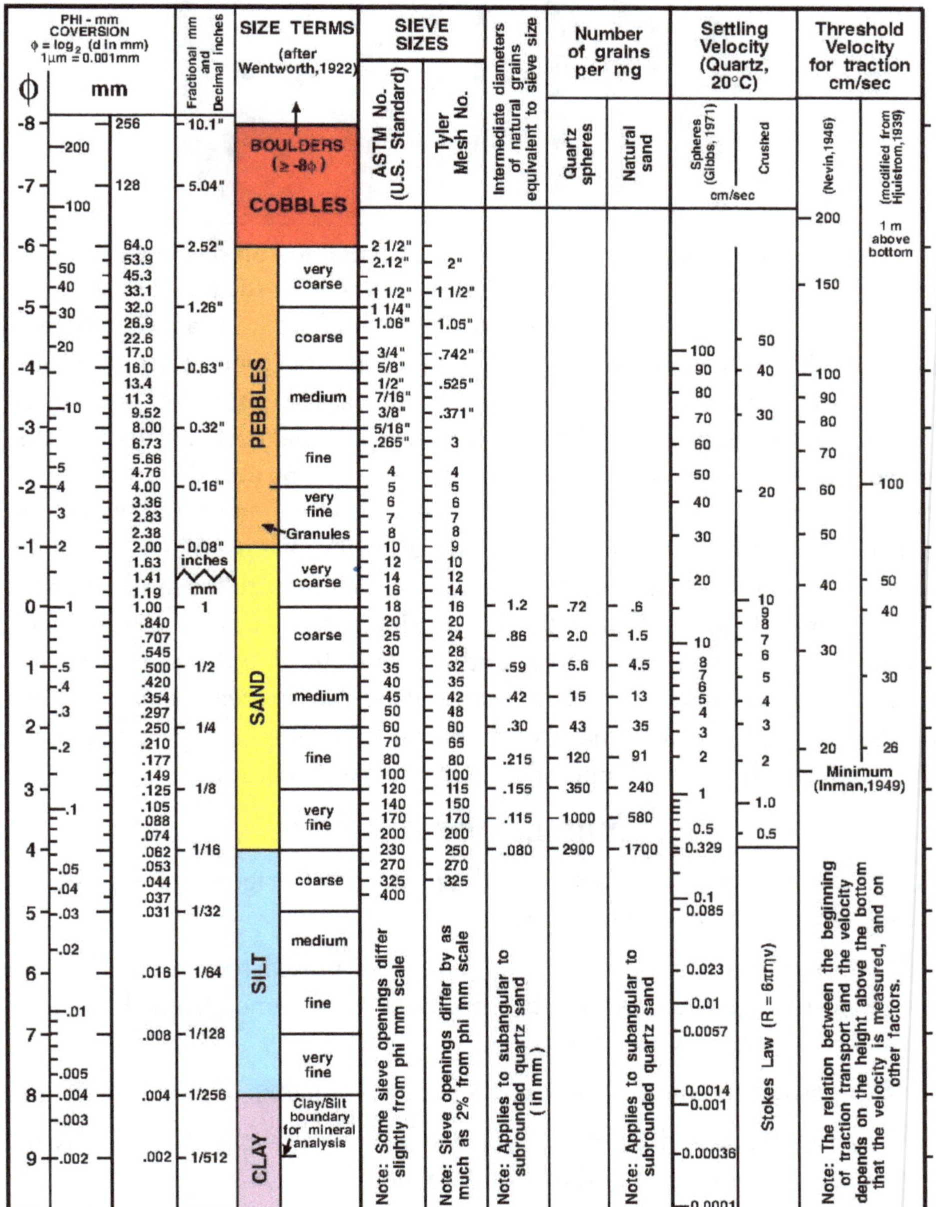

Fig. 3. Wentworth grain size chart for classifying sediments

Sorting is another way to categorize sediments. Sorting refers to how uniform the particles are in terms of size (Fig. 4). If all of the particles are of a similar size, such as in beach sand, the sediment is well-sorted. If the particles are of very different sizes, the sediment is poorly sorted, such as in glacial deposits.

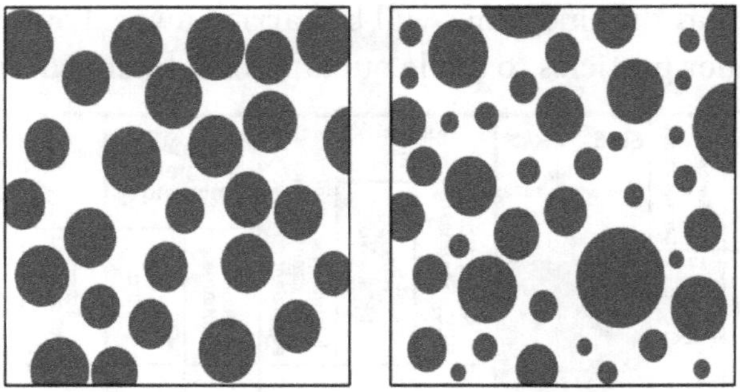

Fig. 4. Well-sorted sediments (left) have particles that are all of a similar size. Poorly sorted sediments (right) consist of particles of a wide range of sizes

Finally, sediment texture can be described based on its **maturity**, or how long the particles have been transported by water or other vectors. Maturity can be reflected by the degree of rounding of the particles, the amount of sorting, and the composition of the sediment. In the case of **rounding**, the more mature the sediment, the rounder the particles, as a result of the particles being abraded over time. A high degree of sorting indicates maturity, because over time the smaller particles will be washed away, and a given amount of energy will move particles of a similar size over the same distance. Lastly, the older and more mature a sediment the higher the quartz content, at least in sediments derived from rock particles. Quartz is a common mineral in terrestrial rocks, and it is very hard and resistant to abrasion. Over time, particles made from other materials are worn away, leaving only quartz behind. Beach sand is a very mature sediment; it is composed primarily of quartz, and the particles are rounded and of similar size (well-sorted).

SOURCES OF SEDIMENTS

Sediments are also classified based on their source of origin. There are four main categories for the origin of marine sediments:

- **Lithogenous** sediments are derived from preexisting rock. They are also called **terrigenous** sediments since most of it comes from the land masses and makes its way into the ocean.
- **Biogenous** sediments are composed of the remains of marine organisms.
- **Hydrogenous** sediments are formed when materials that are dissolved in water precipitate out and form solid particles.
- **Cosmogenous** sediments are derived from extraterrestrial sources.

HOW ARE MARINE SEDIMENTS COLLECTED?

Corers

Corers work by boring a large tube into the benthos and then bringing up a column, or core, of sediment intact within the tube. Caps can automatically seal off the ends of the core after it has pulled up a sample, protecting the sample and keeping it intact. Different sizes and approaches work with different organisms and sediment types.

The gravity corer

Gravity corers are widely used for the collection of the smallest marine metazoans (meiofauna) from subtidal grounds.

The gravity corer is basically a weighted tube mounted within a frame that descends by gravity from the research vessel to the sea floor, where it penetrates the sediment to a given depth, filling the tube with sediment in the process. The hydraulically-damped gravity corer has a slow rate of penetration that is controlled by a water-filled piston. Disturbance of the water-sediment interface is minimal and maximum penetration in mud is about 70 cm. A core-catcher on the bottom of the tube moves into place when retrieval begins, trapping the sediment sample in the corer. A winch, onboard the ship, slowly brings the corer back to the surface.

Fig. 5. Gravity corer

Multiple corer

To test local faunal variations, it is necessary to recover several discrete samples from a single station. The samples could be obtained by multiple casts with a single core tube, but this approach requires much valuable ship time. A multiple corer incorporates the separate coring tubes into a single core body. It consists of a system to which a series of tubes measuring about 4 cm in diameter are attached. Above the system, a weight is mounted and this falls down onto the assembly system when the multicorer touches the sediment. The falling weight drives the tubes into the sea bed so that when they are raised again each of them contains a drilling core with sediment from the sea floor. When studying the benthic communities, scientists then take account of the depth at which these tiny creatures were found below the sea floor.

Fig. 6. Multiple corer

Fig. 7. Reineck boxcorer

Reineck boxcorer

The boxcorer takes relatively undisturbed samples. The equipment operates by a self releasing trigger system triggered by the frame touching the sea bed. The square box is pushed into the bottom by gravity force of the weight mounted on the top of the box retainer. A spade freed by the trigger-mechanism closes the sample box during the recovery of the unit preventing the sample being washed-out.

Box corers

SMBA box corer

The SMBA (Scottish Marine Biological Association) box corer is designed to take an undisturbed sediment sample of 600mm square, up to a maximum depth of around 450mm. The corer consists of a sample box mounted on gimbals, and a spade assembly. The box corer is lowered onto the seabed on a wire at a controlled rate until its frame rests on the bottom. The sample bucket is forced into the sediment by the weight of the corer. As the corer is slowly pulled out of the sediment, a mechanism allows the spade to swing below the sample box sealing in the sediment. Simultaneously, spring loaded flaps above the sample box are closed to prevent the sample being disturbed during recovery. The use of this equipment is very dependent on the type of sediment to be sampled and the current sea state.

NIOZ (Haja) corer

The NIOZ (Koninklijk Nederlands Instituut voor Onderzoek der Zee, the Royal Netherlands Institute for Sea Research) box corer works on a similar principle to the SMBA Box Corer. It is able to take undisturbed samples of 500mm diameter or 500mm square, up to 500mm depth. It differs in design to the SMBA corer in that it has two spades, which operate from either side of the sample box. This design results in a more balanced 'pull-out' of the corer, reducing the risk of disturbing the sediment during sampling.

Day Grab

The Day Grab is a small corer consisting of two quadrant-shaped jaws mounted on a frame. Removable lead weights are bolted to the frame to allow the weight of the corer to be adjusted to suit the sediment type. The Day Grab is lowered to the sea floor on a wire with its two jaws locked open. When the grab contacts the sea floor, a trigger mechanism unlocks the jaws. The action of pulling the wire to recover the grab causes the jaws to rotate towards each other and close to grab a sample of sediment.

Dredging

Dredging gathers loose rocks sitting on the ocean floor using a technique that has changed little in hundreds of years. It is still a useful way of mapping the broad-scale distribution of rock types on the sea floor.

Dredgers have a chain-link bag with large metal-jawed opening that scoops the contents into the bag. The dredging equipment is lowered to the seabed on a cable and dragged along the bottom for some distance before being brought to the surface.

Performing a dredge

The rock dredge package is lowered to the sea floor using a steel wire cable. This means that there is no data transmission, such as video, to show what is happening below. Depending on the ocean depth, the amount of cable deployed to undertake the transit can be up to two and a half times the water depth.

When the rock dredge package and anchor weight chain reach the sea floor initially a reduction in the cable loading, of approximately one tonne, will be observed on the winch monitoring system. The cable loading will then be seen to increase as the dredge is dragged over the sea floor. The cable monitoring system will show 'spike' loadings as the rock dredge picks up samples or snags in the topography. This will continue until the transit line is completed and the rock dredge is recovered to the vessel.

During the dredging transit line, the speed of the research vessel controlling the dredge is kept between 0.5 to 1.5 knots. The speed of the vessel is adjusted as necessary depending on the loadings experienced on the winch monitoring system of the vessel. This means it takes many hours to perform a dredge and many more before the crew and scientists find out if it has been successful.

How the dredge works

The rock dredging equipment comprises of a mild steel frame with an opening approximately 1m wide by 0.5m high, and 0.25m breadth. Attached to the back of the frame, and secured by wire and shackles, is a steel net approximately 1.5m long. Secured to the net, by chain and shackles, is a 0.3m diameter by 0.8m long pipe 'bucket' used as a weight. The Dredge is deployed connected to the vessel's main trawl warp (steel cable), nominally 14,700 metres in length, by an anchor chain and weak link.

The anchor chain acts as weight to keep the rock dredge on the sea floor during the dredge line transit. The rough terrain causes the frame to bounce around, so the anchor at the front and 'bucket' weight at the back keep it grounded. The weak links are a system safety device used to separate the main warp from a trapped rock dredge package.

A strangle wire, wrapped through the chain mail and also connected via a weak link to the main warp enables the dredge to be recovered in the event of snagging. The strangle wire works by closing the 'neck' of the rock dredge net and pulling the dredge net over the steel frame. This tumbling action hopefully frees the rock dredge from its trapped position without spilling the contents.

During transit lines (dredging) where the sea floor topography is known, or expected to be, extremely broken or rough a sacrificial pennant wire is attached. It runs, approximately 500 metres, between the rock dredge package and the vessels main trawling warp to prevent damage to the vessels trawling warp. An acoustic pinger can be secured to the trawling warp/pennant wire 100–300 metres from the rock dredge package. The acoustic pinger can help to indicate that the rock dredge package is on the sea floor by the separation distance of the pinger and the sea floor being less than the length of cable between the pinger and dredge net.

Dredges have been carried out on a number of research expeditions undertaken by the NOC, enabling us to discover more about the sea floor in a range of oceanic locations. Our dredging equipment is available for use via the National Marine Equipment Pool.

Trawling

Trawling usually happens in mid-water or at the bottom of the sea, and is used for sampling mega-fauna including fish, sea urchins and sea cucumbers. Trawls have a cone shaped net attached to a metal frame, and are towed on a wire behind a ship. Once a trawl has reached the correct depth its net is opened and samples are taken.

Performing a trawl

Before trawling begins it important to know if the terrain is suitable as extruding topographic features such as underwater mountains, cables and other obstacles need to be avoided. If the area is not well mapped a survey of the area will be performed first using echo sounders on board the ship or by deploying a Towed Ocean Bottom Instrument (TOBI). Once the trawl is in progress, the crew cannot visually see what is happening, as there are no cameras; instead they will use acoustic transmissions to give depth and angle information.

Trawling is a slow process. As the wire is let out it billows, so many extra metres of wire is needed to get to the right depth. For example, in 5,000m of water depth around 13,000m of wire will be required. Consequently, the trawl will be many kilometres behind the ship and many kilometres down, making both guiding and steering difficult.

The NOC mainly use two different kinds of trawl for scientific research, which are detailed below.

Agassiz trawl

- Used with a clump weight can be fished almost vertically, which is ideal for confined spaces such as canyons
- Small, basic trawl that is easy to use
- Wire out to depth ratio 2:1

- Time at the bottom is up to one hour before the net is full.

Otter Trawl Semi-Balloon 14

- A commercial shrimp trawl, used scientifically to carry out a bottom trawl
- Highly specialised and requires both scientists and technicians to operate
- Wire out to depth ratio 3:1 with 14-metre ground line
- Originally used on the Challenger research vessel, allowing sample comparisons to be made with previous years.

SURFACE SEDIMENT SAMPLES FROM EARLY AGE OF SEAFLOOR EXPLORATION

Late nineteenth and early twentieth century oceanographic expeditions set out to explore the vast and then widely unknown deep ocean. The voyage of HMS *Challenger* is a notable example. As she sailed around the globe between 1872–76, researchers mapped for the first time the shape of the ocean basins and described over 4500 new species of marine life. These early expeditions have important historical significance, as they mark the beginning of modern oceanography and stimulated further ocean exploration.

From a scientific perspective, the observations and material acquired by these historical expeditions have great potential for global change research, as they provide a pre-1900 baseline of the marine environment. Yet historical seafloor sediment samples remain largely underutilized, because early seafloor sampling techniques involved collecting surficial sediments with instruments like the sounding tube, dredge or even the anchor. All these instruments can penetrate below the surface and disturb the top layer of the sediment. As a result, such historical sediment samples might contain surface (Holocene) sediments mixed with deeper, glacial material, potentially hindering their use as a historical baseline of the modern marine environment. Coring techniques provide more accurate sediment chronology; however, historical samples represent the seafloor environment as much as 50 years earlier than the earliest core samples collected, and thus contain sediments without any objects deposited after 1900. These uncontaminated historical samples can be useful for chemical analyses of the seafloor (e.g., pollution trends), single-specimen analysis and investigations of species range shifts and invasions in the past century. Therefore, it is important to assess the degree to which historical sediment samples represent Holocene or mixed-Pleistocene sediments.

One way to assess the degree of glacial mixing in the historical material would be to determine its absolute age using the radiocarbon dating technique, or to use glacial material proxies (e.g., Mg/Ca, oxygen isotopes). However, in cases where the extent of mixing is small, the exponentially decaying nature of the radiocarbon analysis can cause an ambiguous dating, and the isotopic analysis would require a large enough number of specimens to correctly represent the extent of the glacial mixing. Here we propose a complimentary method that uses planktonic foraminifera assemblage composition as a climate-sensitive fingerprint of the sediment age. Planktonic foraminifera (PF) are single-celled zooplankton that produce calcium carbonate shells and, upon death, accumulate in great numbers on the ocean floor. PF assemblage composition is sensitive to sea-surface temperature and its change between glacial and interglacial times has been used to determine the magnitude of glacial ocean cooling.

In this brief report, we make use of the temperature sensitivity of PF and compare the composition of their assemblages in nine historical (> 100 years old) samples against reference PF assemblages from the Holocene (Siccha and Kucera, 2017) and the Last Glacial Maximum (Kucera et al., 2005a). We test whether it is possible to recover the extent of glacial mixing in historical seafloor sediment samples using PF assemblage composition. This new biometric method contributes to a more multidisciplinary approach to dating historical sediments.

Historical Samples

Historical samples were retrieved from the Ocean-Bottom Deposits (OBD) Collection held by The Natural History Museum in London. The OBD Collection holds about 40,000 historical samples from all the world's oceans, including most of the sediment samples collected by HMS *Challenger* and the British Royal Navy survey ships. The OBD samples are kept sealed in their original glass jars and tubes and are usually dry as the result of the long (over 100 years) storage. We selected nine samples collected between 1874 and 1905, chosen to cover different oceans, latitudes and historical marine expeditions. Half of the amount available in the OBD containers was further split into two equal parts, leaving an archive sample and a sample to be processed. The sample processing consisted of weighing, wet washing over a 63 μm sieve and drying in a 60°C oven. The residues were further dry sieved over a 150μm sieve and the coarser fraction was split with a microsplitter as many times as needed to produce a representative aliquot containing around 300 PF shells. All PF specimens in each of the nine final splits were picked, glued to a micropaleontology slide and identified under a stereomicroscope to species level, resulting in a total of 2,611 individuals belonging to 31 species.

Holocene and Last Glacial Maximum Data

We tested whether the composition of PF assemblages in the historical samples is more similar to assemblages of the Holocene (last 11,700 years) or the Last Glacial Maximum (LGM, 21,000 years ago). The Holocene census dataset (i.e., marine surface sediment samples taken by coring methods after 1945) was recently curated and published as the ForCenS dataset, comprising 4205 assemblage counts from unique sites. Three LGM datasets from the MARGO project were merged following the taxonomic standardization. This merged LGM dataset includes 1165 counts from 389 unique sites. Moreover, local estimates of open-ocean sedimentation rates are available for 156 samples in the LGM datasets, and were used to analyse the results.

The assemblage compositions of the nine historical samples were then compared to the samples from the geographically nearest site in the Holocene and LGM datasets. The distances between sites were calculated using the World Geodetic System of 1984. We then compiled annual mean and standard deviation values of sea surface temperature (SST) from the World Ocean Atlas 2013 (WOA13, 0 meters depth) for each of the 27 sites (historical, Holocene and LGM) to evaluate whether the neighboring sites are at similar SST ranges. Most neighboring sites had similar SST values except the two most southern samples. The exceptional Holocene nearest sample neighboring M.8780 was substituted by the fourth nearest neighbor, 66 km farther but more similar in SST. The mean distance between our nine sites and their nearest neighbor set was 253 km in the Holocene data and 415 km in the LGM data.

Compositional Similarity

Assemblage similarity was expressed using the Morisita-Horn index, which is an abundance-based overlap measure that preserves essential properties of similarity measurements. The Morisita-Horn calculates the compositional similarity by pairwise comparison of the relative abundance of each species, and is robust to under-sampling (i.e., rare species occurrence). The index was calculated using the under-sampling bias correction, bootstrap confidence intervals based on 100 replicates and the R package *SpadeR* (version 0.1.1).

For each of the nine historical assemblages, we calculated the Morisita-Horn index three times: between (1) historical and neighboring Holocene assemblages, (2) historical and neighboring LGM assemblages and (3) neighboring Holocene and LGM assemblages. This third comparison gives us a baseline index value of how much the PF assemblage composition changed locally since the LGM. If historical samples are representative of surface sediments, the similarity index calculated between historical and Holocene samples should be higher than between historical and LGM samples, with non-overlapping confidence intervals. However, if historical samples are a mixture of Holocene and LGM material, confidence intervals of the historical-Holocene and historical-LGM comparisons overlap. Confidence intervals might also overlap if the baseline index value calculated between the Holocene and the LGM reference samples is high. A high baseline value means that the local Holocene and LGM assemblages are similar and thus there is less statistical potential to detect whether a historical sediment is a surface or mixed-glacial sample. To understand where our biometric methodology would be most diagnostic, we compiled a world map of species turnover since the LGM, by calculating the Morisita-Horn index for each LGM sample and its nearest Holocene neighbor. Compositional similarity indexes were averaged per site. Distances between the Holocene sites and the 389 LGM sites were calculated using the WGS84, and averaged 52 km (median 1.5 km).

LITHOGENOUS SEDIMENTS

Lithogenous or terrigenous sediment is primarily composed of small fragments of preexisting rocks that have made their way into the ocean. These sediments can contain the entire range of particle sizes, from microscopic clays to large boulders, and they are found almost everywhere on the ocean floor. Lithogenous sediments are created on land through the process of weathering, where rocks and minerals are broken down into smaller particles through the action of wind, rain, water flow, temperature- or ice-induced cracking, and other erosive processes. These small eroded particles are then transported to the oceans through a variety of mechanisms:

Fig. 8. River discharge in the Yukon Delta, Alaska. The pale color demonstrates the large amounts of sediment released into the ocean via the rivers

- **Streams and rivers**: Various forms of runoff deposit large amounts of sediment into the oceans, mostly in the form of finer-grained particles (Fig. 8). About 90% of the lithogenous sediment in the oceans is though to have come from river discharge, particularly from Asia. Most of this sediment, especially the larger particles, will be deposited and remain fairly close to the coastline, however, smaller clay particles may remain suspended in the water column for long periods of time and may be transported great distances from the source.

- **Wind**: Wind-borne (**aeolian**) transport can take small particles of sand and dust and move them thousands of kilometers from the source. These small particles can fall into the ocean when the wind dies down, or can serve as the nuclei around which raindrops or snowflakes form. Aeolian transport is particularly important near desert areas.
- **Glaciers and ice rafting**: As glaciers grind their way over land, they pick up lots of soil and rock particles, including very large boulders, that get carried by the ice. When the glacier meets the ocean and begins to break apart or melt, these particles get deposited. Most of the deposition will happen close to where the glacier meets the water, but a small amount of material is also transported longer distances by rafting, where larger pieces of ice drift far from the glacier before releasing their sediment.
- **Gravity**: Landslides, mudslides, avalanches, and other gravity-driven events can deposit large amounts of material into the ocean when they happen close to shore.
- **Waves**: Wave action along a coastline will erode rocks and will pull loose particles from beaches and shorelines into the water.
- **Volcanoes**: Volcanic eruptions emit vast amounts of ash and other debris into the atmosphere, where it can then be transported by wind to eventually get deposited in the oceans (Fig. 9).

Fig. 10. Eruption of the Mayon Volcano in 1984. Much of the material spewed from a volcanic eruption may eventually make its way into the oceans

- **Gastroliths**: An interesting, although relatively minor avenue for the transport of lithogenous sediments to the ocean is in the form of gastroliths. Gastrolith means "stomach stones" and comes from the fact that many animals, including seabirds, pinnipeds, and some crocodiles will deliberately swallow stones in one area and regurgitate them in another. Often these stones swallowed on land will be regurgitated at sea. Why swallow stones?

Possible explanations include using the stones to help grind up food in the stomach, to act as ballast to aid in buoyancy regulation, or to fill the stomach and reduce feelings of hunger during fasting periods on shore.

Most of these processes deposit lithogenous sediment fairly close to shore. The sediment particles can then be transported farther away by waves and currents, where they may eventually escape the continental shelf and reach the deep ocean floor.

Composition

Lithogenous sediments usually reflect the composition of whatever materials they were derived from, so they are dominated by the major minerals that make up most terrestrial rock. This includes quartz, feldspar, clay minerals, iron oxides, and terrestrial organic matter. Quartz (silicon dioxide, the main component of glass) is one of the most common minerals found in nearly all rocks, and it is very resistant to abrasion, so it is a dominant component of lithogenous sediments, including sand.

CHARACTERISTICS OF BIOGENOUS SEDIMENT

Biogenous sediments (*bio* = life, *generare* = to produce) are sediments made from the skeletal remains of once-living organisms. These hard parts include a wide variety of particles such as shells of microscopic organisms (called **tests**), coral fragments, sea urchin spines, and pieces of mollusc shells.

The most important type of biogenous sediment comes from the tests of one-celled microscopic algae and protozoans living in the surface waters of the oceans. When these tests comprise greater than 30% of the particles in the sediment, the sediment is called an **ooze.** Oozes are most abundant beneath open ocean areas where nutrients are available to enhance productivity. In these areas, oozes accumulate at an average rate of 1 centimeter (0.4 inch) per 1000 years. Oozes are generally absent on the continental margins where lithogenous sediments dominate.

Macroscopic sediments contain large remains, such as skeletons, teeth, or shells of larger organisms. This type of sediment is fairly rare over most of the ocean, as large organisms don't die in enough of a concentrated abundance to allow these remains to accumulate. One exception is around coral reefs; here there is a great abundance of organisms that leave behind their remains, in particular the fragments of the stony skeletons of corals that make up a large percentage of tropical sand.

Microscopic sediment consists of the hard parts of microscopic organisms, particularly their shells, or **tests.** Although very small, these organisms are highly abundant and as they die by the billions every day their tests sink to the bottom to create biogenous sediments. Sediments composed of microscopic tests are far more abundant than sediments from macroscopic particles, and because of their small size they create fine-grained, mushy sediment layers. If the sediment layer consists of at least 30% microscopic biogenous material, it is classified as a biogenous **ooze.** The remainder of the sediment is often made up of clay.

The primary sources of microscopic biogenous sediments are unicellular algaes and protozoans (single-celled amoeba-like creatures) that secrete tests of either calcium carbonate ($CaCO_3$) or silica (SiO_2). Silica tests come from two main groups, the **diatoms** (algae) and the **radiolarians** (protozoans) (Fig. 10).

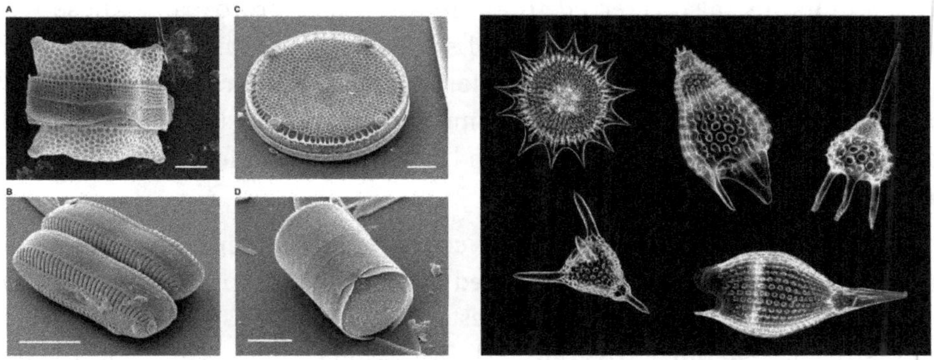

Fig. 10. Various diatom (left) and radiolarian (right) tests

Diatoms are important members of the phytoplankton, the small, drifting algal photosynthesizers. A diatom consists of a single algal cell surrounded by an elaborate silica shell that it secretes for itself. Diatoms come in a range of shapes, from elongated, pennate forms, to round, or centric shapes that often have two halves, like a Petri dish (Fig. 11 *left*). In areas where diatoms are abundant, the underlying sediment is rich in silica diatom tests, and is called **diatomaceous earth.**

Radiolarians are planktonic protozoans (making them part of the zooplankton), that like diatoms, secrete a silica test. The test surrounds the cell and can include an array of small openings through which the radiolarian can extend an amoeba-like "arm" or pseudopod (Fig. 11 *right*). Radiolarian tests often display a number of rays protruding from their shells which aid in buoyancy. Oozes that are dominated by diatom or radiolarian tests are called **siliceous oozes.**

Like the siliceous sediments, the calcium carbonate, or calcareous sediments are also produced from the tests of microscopic algae and protozoans; in this case the **coccolithophores** and **foraminiferans.** Coccolithophores are single-celled planktonic algae about 100 times smaller than diatoms. Their tests are composed of a number of interlocking $CaCO_3$ plates (coccoliths) that form a sphere surrounding the cell (Fig. 11 *left*). When coccolithophores die the individual plates sink out and form an ooze. Over time, the coccolithophore ooze lithifies to becomes chalk. The famous White Cliffs of Dover in England are composed of coccolithophore-rich ooze that turned into chalk deposits (Fig. 11 *right*).

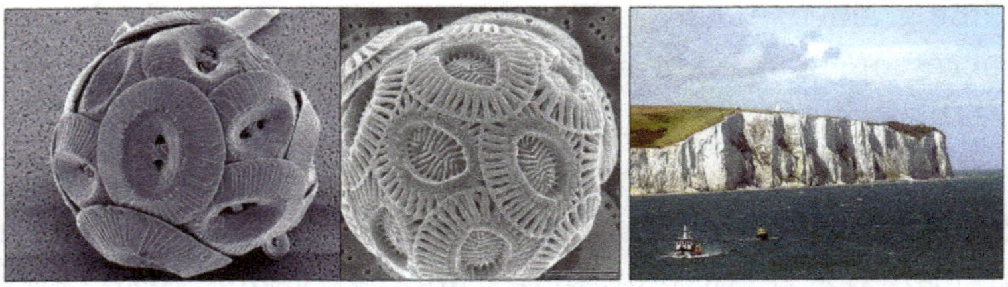

Fig. 11 (Left) coccolithophore tests and (Right) the White Cliffs of Dover

Foraminiferans (also referred to as "forams") are protozoans whose tests are often chambered, similar to the shells of snails. As the organism grows, is secretes new, larger chambers in which to reside. Most foraminiferans are benthic, living on or in the sediment, but there are some planktonic species living higher in the water column. When coccolithophores and foraminiferans die, they form **calcareous oozes.**

Older calcareous sediment layers contain the remains of another type of organism, the **discoasters**; single-celled algae related to the coccolithophores that also produced calcium carbonate tests. Discoaster tests were star-shaped, and reached sizes of 5-40 µm across (Fig. 12). Discoasters went extinct approximately 2 million years ago, but their tests remain in deep, tropical sediments that predate their extinction.

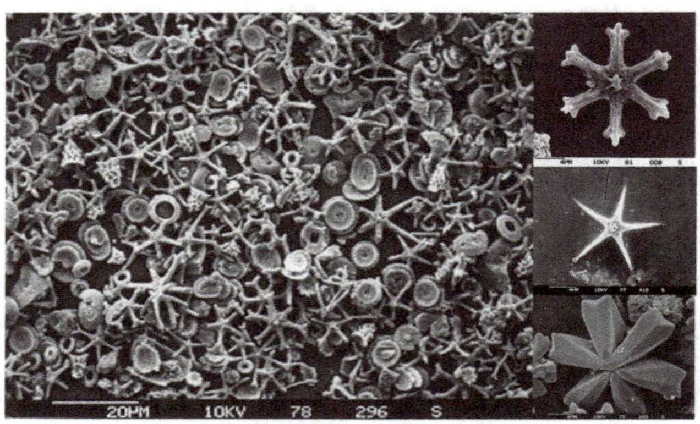

Fig. 12.Discoaster tests. Left: discoster tests with assorted coccoliths. Top right Discoaster surculus; center right: Discoaster pentaradiatus; bottom right: Discoaster surculus

Because of their small size, these tests sink very slowly; a single microscopic test may take about 10-50 years to sink to the bottom! Given that slow descent, a current of only 1 cm/sec could carry the test as much as 15,000 km away from its point of origin before it reaches the bottom. Yet despite this, we find that the sediments in a particular location are well-matched to the types of organisms and degree of productivity that occurs in the water overhead. This means that the sediment particles must be sinking to the bottom at a much faster rate, so that they accumulate below their point of origin before the currents can disperse them. What is the mechanism for this increased sinking rate? Apparently most of the tests do not sink as individual particles; about 99% of them are first consumed by some other organism, and are then aggregated and expelled as large fecal pellets, which sink much more quickly and reach the ocean floor in only 10-15 days. This does not give the particles as much time to disperse, and the sediment below will reflect the production occurring near the surface. The increased rate of sinking through this mechanism is called the "fecal express."

CHARACTERISTICS OF HYDROGENOUS SEDIMENT

Hydrogenous sediments are created from chemical reactions in seawater. Under special chemical conditions, dissolved materials in seawater **precipitate** (form solids). Many types of hydrogenous sediments have economic value.

Hydrogenous sediments include **evaporites**, meaning any type of sediment that forms from the evaporation of seawater. As seawater evaporates, the ions that remain behind can become so concentrated that they will combine with one another to form crystals that precipitate. The two most common types of evaporates are gypsum and halite.

Gypsum is hydrous calcium sulfate ($CaSO_4 \cdot 2H_2O$), and is mined worldwide to make fertilizer, plaster, and cement.

Halite is sodium chloride (NaCl), which is common table salt. When you salt your food, you are eating the evaporated remains of ancient ocean water!

Manganese nodules are another type of hydrogenous sediment. They form marble-size to tennis ball-size lumps of iron and manganese oxide that lie scattered across the deep sea floor where sedimentation rates are particularly low. Although they contain large amounts of manganese, these nodules are economically most important for their cobalt, nickel, and chromium. Their formation is not well understood; however, we do know that they form as concentric layers (like an onion), adding layers of iron and manganese minerals slowly over time. The rates of formation are on the order of 1 to 10 millimeters (0.04 to 0.4 inch) per million years.

Manganese nodules are rounded lumps of manganese and other metals that form on the seafloor, generally ranging between 3-10 cm in diameter, although they may sometimes reach up to 30 cm (Fig. 13). The nodules form in a manner similar to pearls; there is a central object around which concentric layers are slowly deposited, causing the nodule to grow over time. The composition of the nodules can vary somewhat depending on their location and the conditions of their formation, but they are usually dominated by manganese- and iron oxides. They may also contain smaller amounts of other metals such as copper, nickel and cobalt. The precipitation of manganese nodules is one of the slowest geological processes known; they grow on the order of a few millimeters per million years.

For that reason, they only form in areas where there are low rates of lithogenous or biogenous sediment accumulation, because any other sediment deposition would quickly cover the nodules and prevent further nodule growth. Therefore, manganese nodules are usually limited to areas in the central ocean, far from significant lithogenous or biogenous inputs, where they can sometimes accumulate in large numbers on the seafloor (Fig. 13 *right*).

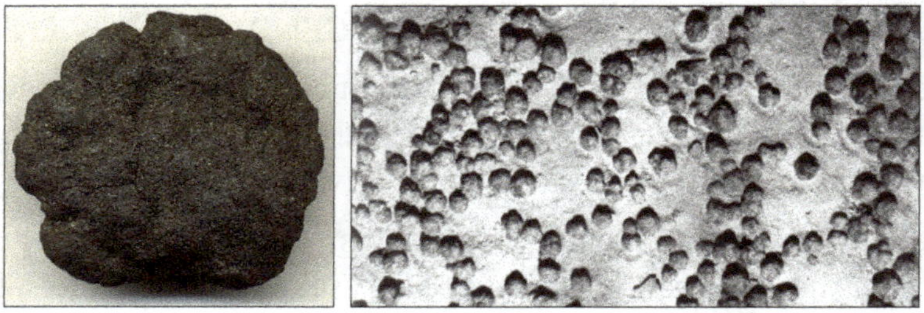

Fig. 13 (Left) manganese nodule from the subtropical eastern Pacific Ocean. The nodule is 4.1 cm in diameter

Because the nodules contain a number of commercially valuable metals, there has been significant interest in mining the nodules over the last several decades, although most of the efforts have thus far remained at the exploratory stage. A number of factors have prevented large-scale extraction of nodules, including the high costs of deep sea mining operations, political issues over mining rights, and environmental concerns surrounding the extraction of these non-renewable resources.

Fig. 14 Oolites

Evaporites are hydrogenous sediments that form when seawater evaporates, leaving the dissolved materials to precipitate into solids, particularly halite (salt, NaCl). In fact, the evaporation of seawater is the oldest form of salt production for human use, and is still carried out today. Large deposits of halite evaporites exist in a number of places, including under the Mediterranean Sea. Beginning around 6 million years ago, tectonic processes closed off the Mediterranean Sea from the

Atlantic, and the warm climate evaporated so much water that the Mediterranean was almost completely dried out, leaving large deposits of salt in its place (an event known as the Messinian Salinity Crisis). Eventually the Mediterranean re-flooded about 5.3 million years ago, and the halite deposits were covered by other sediments, but they still remain beneath the seafloor.

Oolites are small, rounded grains formed from concentric layers of precipitation of material around a suspended particle. They are usually composed of calcium carbonate, but they may also from from phosphates and other materials. Accumulation of oolites results in oolitic sand, which is found in its greatest abundance in the Bahamas (Fig. 14).

Methane hydrates are another type of hydrogenous deposit with a potential industrial application. All terrestrial erosion products include a small proportion of organic matter derived mostly from terrestrial plants. Tiny fragments of this material plus other organic matter from marine plants and animals accumulate in terrigenous sediments, especially within a few hundred kilometers of shore. As the sediments pile up, the deeper parts start to warm up (from geothermal heat), and bacteria get to work breaking down the contained organic matter. Because this is happening in the absence of oxygen (a.k.a. anaerobic conditions), the by-product of this metabolism is the gas methane (CH_4).

Methane released by the bacteria slowly bubbles upward through the sediment toward the seafloor. At water depths of 500 m to 1,000 m, and at the low temperatures typical of the seafloor (close to 4°C), water and methane combine to create a substance known as methane hydrate. Within a few meters to hundreds of meters of the seafloor, the temperature is low enough for methane hydrate to be stable and hydrates accumulate within the sediment.

Methane hydrate is flammable because when it is heated, the methane is released as a gas. The methane within seafloor sediments represents an enormous reservoir of fossil fuel energy. Although energy corporations and governments are anxious to develop ways to produce and sell this methane, anyone that understands the climate-change implications of its extraction and use can see that this would be folly.

CHARACTERISTICS OF COSMOGENOUS SEDIMENT

Cosmogenous sediments originated from outer space. Scientists have used satellites to estimate how much material enters the earth's atmosphere. Current estimates from satellite data suggesting about 100 to 300 tons (mostly cosmic dust) hits earth each day. This is just a tiny fraction of the sediments generated on earth each day. However, early in the history of our Solar System, Earth and other planets, moons, comets and asteroids formed from the gravitational accumulation of extraterrestrial material, but by 4.5 million years ago, most of this cosmogenous accumulation had significantly diminished. However, cosmogenous materials including **iron-nickel** and **stony meteorites** can be found. Although a relatively insignificant source of sediment, meteor fireballs disintegrating in the atmosphere contribute dust that can accumulate measurable amounts in parts of some ocean basins.

Extraterrestrial impacts have changed life on Earth repeatedly, including the mass extinction at the end of the Mesozoic Era associated with the extinction of dinosaurs and many other forms of life on land and in the oceans. **Tektites** are silica glass generated by extraterrestrial impacts: asteroids exploding on the surface and molten material is ejected into the atmosphere where it condenses into a glass-like material.

Cosmogenous sediments come from outer space and have extraterrestrial sources, such as the remains from the impact of large bodies of space material like comets and asteroids. This type of sediment is rarely found by scientists and is the most insignificantly important one.

Origin, Composition and Distribution

The two different types are microscopic spherules and macroscopic meteor debris. Microscopic spherules are made of silicate rock material and were once formed by extraterrestrial impact events that released molten pieces of crust called tektites into space. Other spherules of iron and nickel were created in the asteroid belt and rains down on Earth's surface as part of space dust or micrometeorites. Space dust that falls in oceans eventually dissolves. Meteorite material is referred to as *chondrites* when consisting of silicate rock material and *irons* when consisting of iron and nickel.

Cosmogenous sediments are mainly located near meteor impact structures or can be found in small amounts mixed with a variety of sediments in all kinds of marine environments.

Cosmogenous sediment is fairly rare in the ocean and it does not usually accumulate in large deposits. However, it is constantly being added to through space dust that continuously rains down on Earth. About 90% of incoming cosmogenous debris is vaporized as it enters the atmosphere, but it is estimated that 5 to 300 tons of space dust land on the Earth's surface each day!

PELAGIC AND NERITIC DEPOSITS

The term **neritic** is used to described the shallow part of the ocean near a coast and overlying the continental shelf.

Neritic sediments are generally shallow water deposits formed close to land. They are dominated by lithogenous sources and are typically deposited quickly. **Neritic** sediments cover about ¼ of sea floor and are near landmasses.

The term **pelagic** means "of or relating to the open sea" particularly the upper layers of the ocean away from shore.

Pelagic sediments are generally deep-water deposits mostly oozes (see below) and windblown clays. They are typically finer-grained sediments that are deposited slowly. Because they are deposited far beyond the continental margins they are typically less lithogenous and more biogenous depending on biologic productivity. **Pelagic** sediments cover about ¾ of seafloor and are mostly in deep water.

The distribution of neritic or pelagic sediments is controlled by proximity to sources of **lithogenous sediments** (i.e.: landmasses) and the **productivity** of microscopic marine organisms.

Pelagic Deposits:

Pelagic deposits are comprised of both organic and inorganic sediments and are generally found in the deep oceans in the continental slope, continental rise, trenches, and abyssal plains.

Pelagic Deposits are generally two types:

- Mud
- Ooze

Mud Pelagic deposits:

Muds are very finer sediments and clay. It is largely found in continental slopes and Abbysal plains.

Muds are classified into the following types:

- Red Mud
- Blue Mud
- Green Mud
- Coral Mud

Red Mud:

- Red mud or clay generally originated from Volcanos activities. It contains Iron oxide.

Blue Mud:

- It contains Iron sulfite.

Green Mud:

- It contains Potassium silicate.

Coral Mud:

- It contains a coral reef.

RESOURCES FROM MARINE SEDIMENTS

Energy

The main energy resources associated with marine sediments are petroleum and gas hydrates. The ancient remains of microscopic organisms, buried within marine sediments before they could decompose, are the source of today s petroleum (oil and natural gas) deposits. Petroleum products account for 95% of the economic value of the ocean beds. This mainly includes the oil produced from offshore regions. Today major offshore reserves exist in the Persian Gulf, in the Gulf of Mexico, off Southern California and in the North Sea.

Gas hydrates are unusually compact chemical structures made of water and natural gas. They form only when high pressures squeeze chilled water and gas molecules into an icelike solid. Although hydrates can contain a variety of gases including carbon dioxide, hydrogen sulfide, and larger hydrocarbons such as ethane and propane; methane hydrates are by far the most common hydrates in nature.

Gas hydrates occur beneath Arctic permafrost areas on land and under the ocean floor, where they were discovered in 1976. In deep-ocean sediments, where pressures are high and temperatures are low, water and natural gas combine in such a way that the gas is trapped inside a lattice like cage of water molecules.

Sand and Gravel

The offshore sand and gravel industry is second in economic value only to the petroleum industry. These include the rock fragments that are washed out to sea and shells of marine organisms, is mined by offshore barges using a suction dredge. This material is primarily used as aggregate in concrete, as a fill material in grading projects, and on recreational beaches.

Evaporative Salts

When seawater evaporates, the salts increase in concentration until they can no longer remain dissolved, so they precipitate out of solution and form salt deposits. The most economically useful salts are gypsum and halite i.e, common salt.

Manganese Nodules and Crusts

Manganese nodules are rounded, hard, golf- to tennis-ball-sized lumps of metals that contain significant concentrations of manganese, iron, and smaller concentrations of copper, nickel, and cobalt, all of which have a variety of economic uses.

TRENDS IN OCEAN TEMPERATURE

Measurement of Temperature

Standard type of thermometer is used to measure the surface temperature while reversing thermometers and thermographs are used to measure the subsurface temperature. These thermometers record the temperature up to the accuracy of ±0.02° centigrade.

Layers of Temperature in tropics

Oceans absorb more than 80% of the solar radiation and water which has highest specific heat is the remarkable capacity of storing the heat. The uppermost 10% of the oceans has more heat than the entire atmosphere of earth!

With respect to temperature, there are three layers in the oceans from surface to the bottom in the tropics as follows:

- The first layer represents the top-layer of warm, oceanic water and is 500m thick with temperature ranging between 20° and 25°C. This layer is present within the tropics throughout the year but it develops in mid· latitudes only during summer.
- The thermocline layer represents vertical zone of oceanic water below the first layer and is characterized by rapid rate of decrease of temperature with increasing depth,
- The third layer is very cold and extends upto the deep ocean floor. The polar areas have only one layer of cold water from surface (sea level) to the deep ocean floor.

The radiant energy transmitted from the photosphere of the sun in the form of electromagnetic short waves and received at the ocean surface is called insolation. Besides, some energy, though insignificant, is also received from below the bottom and through the compression of sea water. The amount of insolation to be received at the sea surface depends on the angle of sun's rays, length of day, distance of the earth from the sun and effects of the atmosphere. The mechanism of the heating and cooling of ocean water differs from the mechanism on land because besides horizontal and vertical movements of water, the evaporation is most active over the oceans.

Daily Range of Temperature

The difference of maximum and minimum temperature of a day (24 hours) is known as daily range of temperature. The daily range of temperature of surface water of the oceans is almost insignificant as it is around 1°C only. The daily range of temperature is usually 0.3°C in the low latitudes and 0.2° to 0.3°C in high latitudes.

The diurnal range depends on the

- Conditions of sky (cloudy or clear sky),
- Stability or instability of air and
- Stratification of seawater.

The heating and cooling of ocean water is rapid under clear sky (cloudless) and hence the diurnal range of temperature becomes a bit higher than under overcast sky and strong air circulation. The high density of water below surface water causes very little transfer of heat through conduction and hence the diurnal range of temperature becomes low.

Annual Range of Temperature

The maximum and minimum annual temperatures of ocean water are recorded in August and February respectively in the northern hemisphere. Usually, the average annual range of temperature of ocean water is -12°C but there is a lot of regional variation which is due to regional variation in insolation, nature of seas, prevailing winds, location of seas etc.

Annual range of temperature is higher in the enclosed seas than in the open sea (Baltic Sea records annual range of temperature of 4.4°C or 40°F). The size of the oceans and the seas also affects annual range of temperature e.g., bigger the size, lower the annual range and vice versa. The Atlantic Ocean records relatively higher annual range of temperature than the Pacific Ocean.

CHAPTER

WATER AND SEAWATER 5

INTRODUCTION

The international standard for the properties of fresh water and seawater are specified by the International Association for the Properties of Water and Steam (IAPWS). The properties available include density, viscosity, thermal conductivity, index of refraction, vapour pressure, speed of sound, and surface tension. Those of liquid water are described in IAPWS. For this procedure, only the following properties are provided: density, absolute viscosity, kinematic viscosity, and vapour pressure.

In general, the water properties are a function of temperature (t), pressure (p), and absolute salinity (SA). For fresh water, SA = 0.0. In this procedure, data are provided at standard pressure of 0.101325 MPa and as a function of temperature. The temperature scale is the International Temperature Scale 1990 (ITS-90). At non-standard conditions, water properties should be computed from computer codes described in the following paragraphs.

The values for fresh water were computed via a computer code from NIST (National Institute of Standards and Technology), the National Metrology Institute (NMI) for the United States. Researchers is the manual for the computer code. The sensitivity coefficients are also provided so that the uncertainty in the property may be computed from the uncertainty in temperature per the 25th ITTC procedure on uncertainty analysis. The uncertainties in the IAPWS equations are also provided.

THE DENSITY OF FRESH WATER AND SEAWATER

Fresh water and seawater have very different physical properties. Imagine a freshwater lake in winter. As the air temperature falls the temperature of the water at the surface decreases and its density changes (Fig. 1).

While it is hard to see on the scale in Fig. 1, the maximum density is at 3.98°C. As a lake cools to this temperature in winter the surface waters will sink through convection and warmer water rises to the surface. With continued cooling this warmer surface water also becomes dense and sinks. As the surface water is being cooled the lake will become stratified. That is, the density will increase with depth to 1000 kg m^{-3}, and the deep water will have a temperature of 3.98°C. Imagine the convection continuing until all the water has reached 3.98°C.

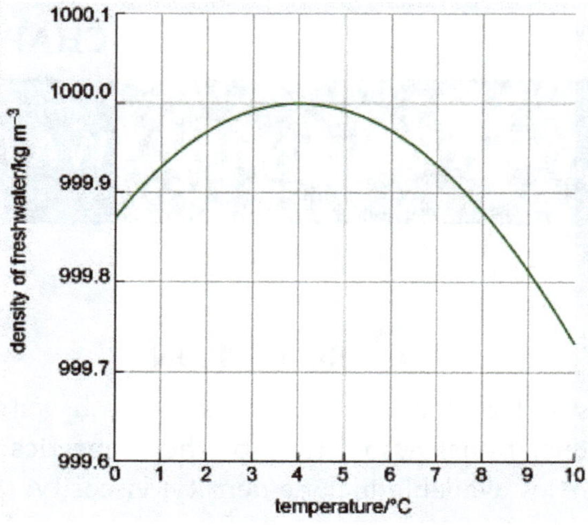

Fig. 1. The density of fresh water plotted against temperature.

Eventually the surface layer of water will be cooled to the freezing point (0°C) and ice will form on the surface. At this point the temperature and density of the water will have a structure like that shown in Fig. 2.

Figure 2b and c show that the temperature and the density increase with depth from the surface to the bottom layer, where the temperature is 3.98 °C and density is at its maximum.

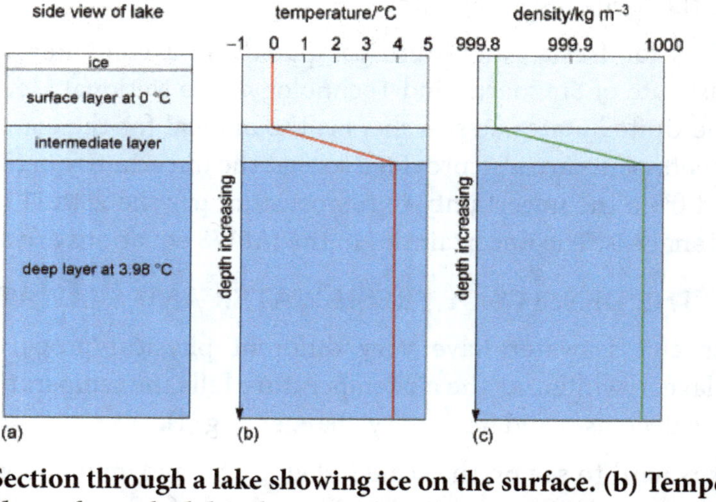

Fig. 2 (a) Section through a lake showing ice on the surface. (b) Temperature structure throughout the lake: there are three layers: a cold surface layer, an intermediate layer and a warm deeper layer. (c) Density structure mirroring the shape of the temperature profile in (b).

Usually when a liquid is heated the molecules acquire more energy and become more widely spaced, so in the same volume, the density decreases. In fresh water the opposite may happen, depending on the temperature. When fresh water at 3.97 °C (Fig. 1) is cooled the density will decrease. You know from the lake (Fig. 2) that ice (i.e. solid water) is less dense and floats. But cooling a liquid usually packs the molecules more closely, which increases the density. This means that below 3.98 °C cooling results in the water molecules spacing out and both the liquid and the solid water below this temperature expand. This is an amazing physical property and is why pipes burst and water in cracks shatters rocks in cold temperatures. The molecular structure of a water molecule is shown in Fig. 3.

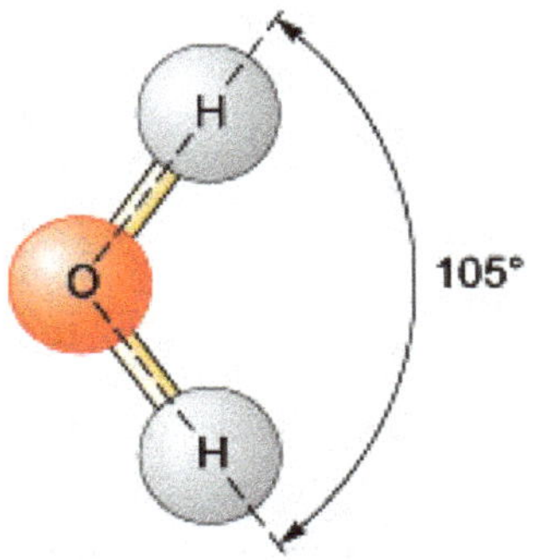

Fig. 3. A molecule of water, H_2O, with the hydrogen and the oxygen atoms sharing electrons.

In H_2O, the oxygen and the hydrogen atoms share electrons, and the angle between the two hydrogen atoms is 105°. This results in a small net negative charge on the oxygen side of the molecule, and a small net positive charge on the hydrogen side. This is a polar structure in which molecules are weakly attracted to each other and form weak 'hydrogen bonds'. At low temperatures a more ordered packing of water molecules develops and the density is reduced. If the temperature is increased but is <3.98 °C, the hydrogen bonds break, but the molecules still pack together closely. Above 3.98 °C, the increase in internal energy means the molecules become more widely spaced and, following Fig. 1, the density decreases.

Seawater is saline and the salt affects the density. Figure 7 shows the vertical structure of the temperature, salinity, and density at latitude 20° S in the Atlantic Ocean over an abyssal plain almost 5300 m deep.

Figure 4a shows 28 °C temperature at the surface to below 1°C at the sea floor. The variation in salinity in 7b clearly affects the temperature of maximum density shown in Fig. 4c. The seawater density ranges from 10^{24} to 10^{28} kg m^{-3}, ~24-28 kg m^{-3} denser than fresh water (Fig. 1), and density increases with depth and the water is stratified all the way to the sea floor (although >3 000 m depth there is only a small increase). Because seawater is typically 1024-1028 kg m^{-3} there is a density anomaly, σ_t (pronounced 'sigma t') given by:

$$\sigma_t = (\sigma - 1000) \text{ kg m}^{-3} \qquad \text{(Eq. 1)}$$

where σ is the density of water.

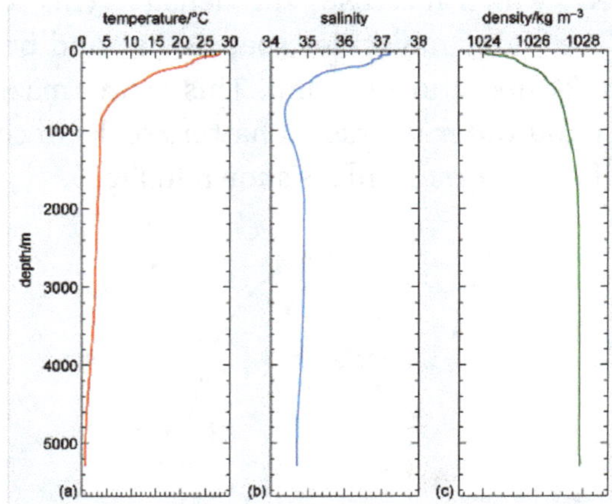

Fig. 4. The physical properties of seawater at 20° S in the Atlantic Ocean against depth. (a) Temperature; (b) salinity; (c) density.

PROPERTIES OF WATER

The most obvious feature of the oceans is that they contain water. Water is so ubiquitous that it may not seem like a very interesting substance, but it has many unique properties that impact global oceanographic and climatological processes. Many of these processes are due to hydrogen bonds forming between water molecules.

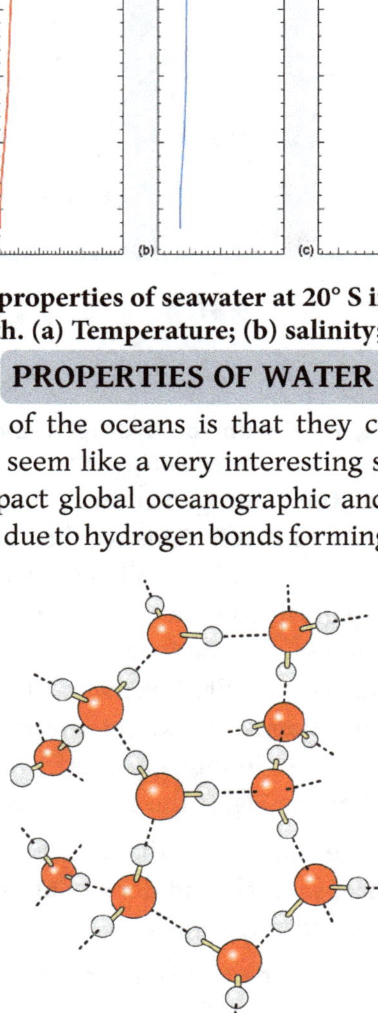

Fig. 5. Hydrogen bonds (dashed lines) between water molecules. Oxygen atoms are shown in red, hydrogen atoms in white

The water molecule consists of two hydrogen atoms and one oxygen atom. The electrons responsible for the bonds between the atoms are not distributed equally throughout the molecule, so that the hydrogen ends of water molecules have a slight positive charge, and the oxygen end has a slight negative charge, making water a polar molecule. The negative oxygen side of the molecule forms an attraction to the positive hydrogen end of a neighboring molecule. This rather weak force of attraction is called a hydrogen bond (Fig. 5). If not for hydrogen bonds, water would vaporize at -68°C, meaning liquid water (and thus life) could not exist on Earth. These hydrogen bonds are responsible for some of water's unique properties:

1. Water is the only substance to naturally exist in a solid, liquid, and gaseous form under the normal range of temperatures and pressures found on Earth. This is due to water's relatively high freezing and vaporizing points (see below).
2. Water has a high heat capacity, which is the amount of heat that must be added to raise its temperature. Specific heat is the heat required to raise the temperature of 1 g of a substance by 1°C. Water has the highest specific heat of any liquid except ammonia (Table 1).

Table 1 Specific heat values for a number of common substances

	Specific Heat (calories/g/°C)
Ammonia	1.13
Water	1.00
Acetone	0.51
Grain Alcohol	0.23
Aluminum	0.22
Copper	0.09
Silver	0.06

Water is therefore one of the most difficult liquids to heat or cool; it can absorb large amounts of heat without increasing its temperature. Remember that temperature reflects the average kinetic energy of the molecules within a substance; the more vigorous the motion, the higher the temperature. In water, the molecules are held together by hydrogen bonds, and these bonds must be overcome to allow the molecules to move freely. When heat is added to water the energy must first go to breaking the hydrogen bonds before the temperature can begin to rise. Therefore, much of the added heat is absorbed by breaking H bonds, not by increasing the temperature, giving water a high heat capacity.

Hydrogen bonds also give water a high latent heat; the heat required to undergo a phase change from solid to liquid, or liquid to gas. The latent heat of fusion is the heat required to go from solid to liquid; 80 cal/g in the case of ice melting to water. Ice is a solid because hydrogen bonds hold the water molecules into a solid crystal lattice. As ice is heated, the temperature rises up to 0°C. At that point, any additional heat goes to melting the ice by breaking the hydrogen bonds, not to increasing the temperature. So as long as ice is present, the water temperature will not increase. This is why your drink will remain cold as long as it contains ice; any heat absorbed goes to melting the ice, not to warming the drink.

When all of the ice is melted, additional heat will increase the temperature of the water 1°C for each calorie of heat added, until it reaches 100°C. At that point, any additional heat goes to overcoming the hydrogen bonds and turning the liquid water into water vapor, rather than increasing the water temperature. The heat required to evaporate liquid water into water vapor is the latent heat of vaporization which has a value of 540 cal/g (Fig. 6).

The high heat capacity of water helps regulate global climate, as the oceans slowly absorb and release heat, preventing rapid swings in temperature. It also means that aquatic organisms aren't as subjected to the same rapid temperature changes as terrestrial organisms. A deep ocean organism may not experience more than a 0.5°C change in temperature over its entire life, while a terrestrial species may encounter changes of more than 20°C in a single day!

Water dissolves more substances than any other liquid; it is a "universal solvent", which is why so many substances are dissolved in the ocean. Water is especially good at dissolving ionic salts; molecules made from oppositely charged ions such as NaCl (Na^+ and Cl^-). In water, the charged ions attract the polar water molecules. The ions get surrounded by a layer of water molecules, weakening the bond between the ions by up to 80 times. With the bonds weakened between ions, the substance dissolves (Fig. 7).

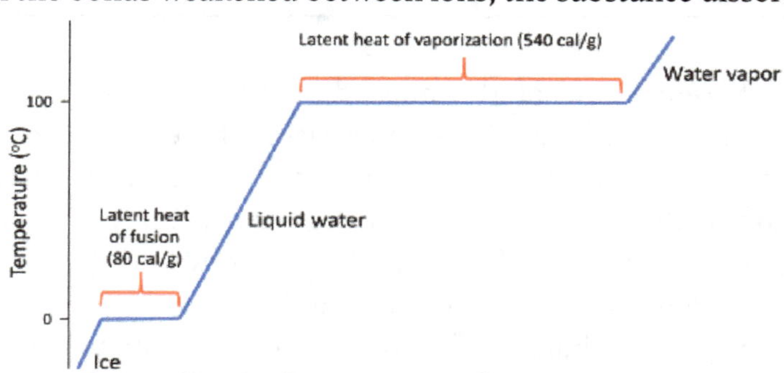

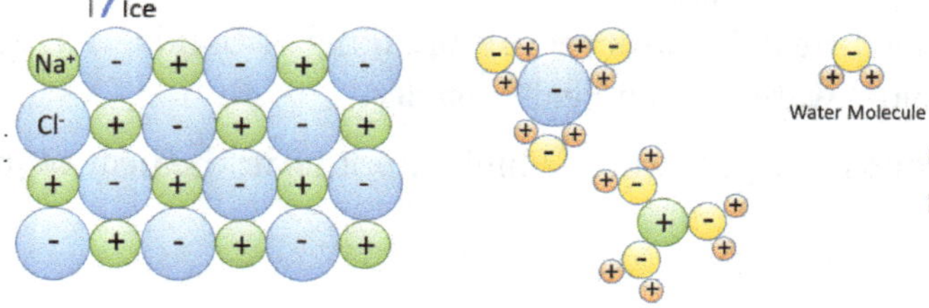

Fig. 6. Latent … **f fusion is the**
heat required to … **e heat required**

Fig. 7. Attraction between polar water molecules and charged ions (such as in NaCl) is greater than the attraction between the charged ions, causing the ions to dissociate and the salt to dissolve

The solid phase is less dense than the liquid phase. In other words, ice floats. Most substances are denser in the solid form than in the liquid form, as their molecules are more closely packed together as a solid. Water is an exception: the density of fresh water is 1.0 g/cm^3, while the density of ice is 0.92 g/cm^3, and once again, this is due to the action of hydrogen bonds.

As water temperature cools the molecules slow down, eventually slowing enough that hydrogen bonds can form and hold the water molecules in a crystal lattice. The molecules in the lattice are spaced farther apart than the molecules in liquid water, which makes ice less dense than liquid water (Fig. 8). This is familiar to anyone who has ever left a full water bottle in the freezer, only to have it burst as the water freezes and expands.

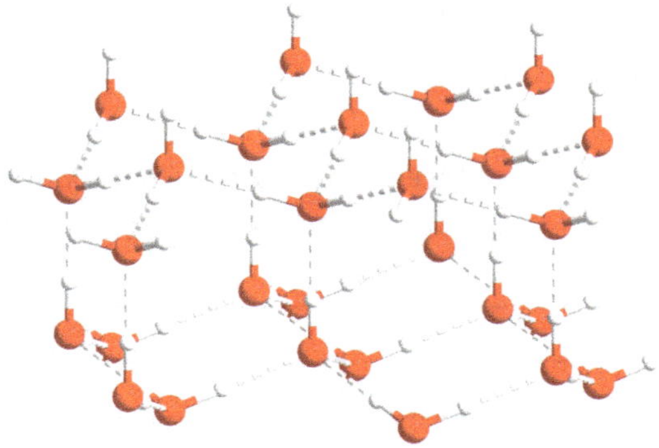

Fig. 8. Crystal lattice structure of ice, showing water molecules held together by hydrogen bonds

But the relationship between temperature and water density is not a simple linear one. As water cools, its density increases as expected, as the water molecules slow down and get closer together. However, fresh water reaches its maximum density at a temperature of 4°C, and as it cools beyond that point its density declines as the hydrogen bonds begin to form and the intermolecular spacing increases. The density continues to decline until the temperature reaches 0°C and ice crystals form, reducing the density dramatically.

There are a number of important implications to ice being less dense than water. Ice floating on the surface of the ocean helps regulate ocean temperatures, and therefore global climate, by influencing the amount of sunlight that is reflected rather than absorbed. On a smaller scale, surface ice can prevent lakes and ponds from freezing solid during the winter. As fresh surface water cools, the water gets denser, and sinks to the bottom. The new surface water then cools and sinks, and the process is repeated in what is referred to as overturning, with denser water sinking and less dense water moving to the surface only to be cooled and sink itself. In this way, the entire body of water is cooled somewhat evenly. This process continues until the surface water cools below 4°C.

Below 4°C, the water becomes *less* dense as it cools, so it no longer sinks. Instead, it remains as the surface, getting colder and less dense, until it freezes at 0°C. Once fresh water freezes, the ice floats and insulates the rest of the water beneath it, reducing further cooling. The densest bottom water is still at 4°C, so it does not freeze, allowing the bottom of a lake or pond to remain unfrozen (which is good news for the animals living there) no matter how cold it gets outside.

The dissolved salts in seawater inhibit the formation of the crystal lattice, and therefore make it harder for ice to form. So seawater has a freezing point of about -2°C (depending on salinity), and freezes before a temperature of maximum density is reached. Thus seawater will continue to sink as it gets colder, until it finally freezes.

Water has a very high surface tension, the highest of any liquid except mercury (Table 2).

Table 2 Surface tensions of various liquids

Liquid	Surface Tension (millinewton/meter)	Temperature °C
Mercury	487.00	15
Water	71.97	25
Glycerol	63.00	20
Acetone	23.70	20
Ethanol	22.27	

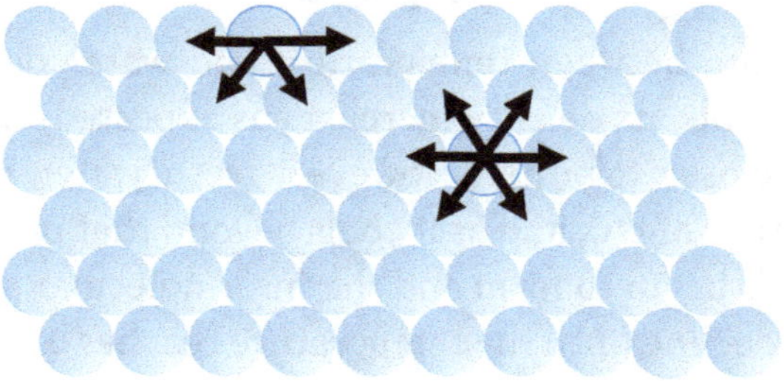

Fig. 9. The net attractive force between molecules at the surface is inwards, leading to surface tension. For molecules in the center, the force is equal in all directions

Water molecules are attracted to each other by hydrogen bonds. For molecules not at the water surface, they are surrounded by other water molecules in all directions, so the attractive forces are evenly distributed in all directions. But for molecules at the surface there are few adjacent molecules above them, only below, so all of the attractive forces are directed inwards, away from the surface (Fig. 9). This inwards force is what causes water droplets to take on a spherical shape, and water to bead up on a surface, as the spherical shape provides the minimum possible surface area. These attractive forces also cause the surface of the water to act like an elastic "skin" which allows things like insects to sit on the water's surface without sinking.

CHEMICAL PROPERTIES

Acid-base reactions

Water undergoes various types of chemical reactions. One of the most important chemical properties of water is its ability to behave as both an acid (a proton donor) and a base (a proton acceptor), the characteristic property of amphoteric substances. This behaviour is most clearly seen in the autoionization of water:

$H_2O(l) + H_2O(l)\ H_3O^+(aq) + OH^-(aq)$,

where the (l) represents the liquid state, the (aq) indicates that the species are dissolved in water, and the double arrows indicate that the reaction can occur in either direction and an equilibrium condition exists. At 25 °C (77 °F) the concentration of hydrated H^+ (i.e., H_3O^+, known as the hydronium ion) in water is 1.0×10^{-7} M, where M represents moles per litre. Since one OH^- ion is produced for each H_3O^+ ion, the concentration of OH^- at 25 °C is also 1.0×10^{-7} M. In water at 25 °C the H_3O^+ concentration and the OH^- concentration must always be 1.0×10^{-14}:

$[H^+][OH^-] = 1.0 \times 10^{-14}$,

where $[H^+]$ represents the concentration of hydrated H^+ ions in moles per litre and $[OH^-]$ represents the concentration of OH^- ions in moles per litre.

When an acid (a substance that can produce H^+ ions) is dissolved in water, both the acid and the water contribute H^+ ions to the solution. This leads to a situation in which the H^+ concentration is greater than 1.0×10^{-7} M. Since it must always be true that

$[H^+][OH^-] = 1.0 \times 10^{-14}$ at 25 °C, the $[OH^-]$

must be lowered to some value below 1.0×10^{-7}. The mechanism for reducing the concentration of OH^- involves the reaction

$H^+ + OH^- \rightarrow H_2O$,

which occurs to the extent needed to restore the product of $[H^+]$ and $[OH^-]$ to 1.0×10^{-14} M. Thus, when an acid is added to water, the resulting solution contains more H^+ than OH^-; that is, $[H^+] > [OH^-]$. Such a solution (in which $[H^+] > [OH^-]$) is said to be acidic.

The most common method for specifying the acidity of a solution is its pH, which is defined in terms of the hydrogen ion concentration:

$pH = -\log [H^+]$,

where the symbol log stands for a base-10 logarithm. In pure water, in which

$[H^+] = 1.0 \times 10^{-7}$ M, the pH = 7.0.

For an acidic solution, the pH is less than 7. When a base (a substance that behaves as a proton acceptor) is dissolved in water, the H^+ concentration is decreased so that $[OH^-] > [H^+]$. A basic solution is characterized by having a pH > 7. In summary, in aqueous solutions at 25°C:

neutral solution	[H+] = [OH-]	pH = 7
acidic solution	[H+] > [OH-]	pH < 7
basic solution	[OH-] > [H+]	pH > 7

PROPERTIES OF SEA WATER

Seawater properties are different from pure water seawater has the following properities.

1. High density and specific gravity
2. Depression in freezing point
3. Elevated boiling point
4. Higher osmotic pressure
5. Good conductor of electricity

Sea water is a mixture of 96.5% pure water and 3.5% other material such as salts, dissolved gases, organic substances, and undissolved particles. Its physical properties are mainly determined by the 96.5% pure water. The physical properties of pure water will therefore be discussed first.

Pure water, when compared with fluids of similar composition, displays most uncommon properties. This is the result of the particular structure of the water molecule H_2O: The hydrogen atoms carry one positive charge, the oxygen atom two negative charges, but the atom arrangement in the water molecule is such that the charges are not neutralized the charges would be neutralized if the angle were 180° rather than 105°.

The major consequences of the molecular structure of pure water are

1. The water molecule is an electric dipole, forming aggregations of molecules (polymers), of on average six molecules at 20°C. Therefore, water reacts slower to changes than individual molecules; for example the boiling point is shifted from -80°C to 100°C, the freezing point from -110°C to 0°C.
2. Water has an unusually strong disassociative power, i.e. it splits dissolved material into electrically charged ions. As a consequence, dissolved material greatly increases the electrical conductivity of water. The conductivity of pure water is relatively low, but that of sea water is midway between pure water and copper. The angle 105° is close to the angle of a tetrahedron, i.e. a structure with four arms emanating from a centre at equal angles (109° 28′).

As a result, oxygen atoms in water try to have four hydrogen atoms attached to them in a tetrahedral arrangement. This is called a “hydrogen bond”, in contrast to the (ionic) molecular bond and covalent bonding. Hydrogen bonds need a bonding energy 10 to 100 times smaller than molecular bonds, so water is very flexible in its reaction to changing chemical conditions.

3. Tetrahedrons are of a more wide-meshed nature than the molecular closest packing arrangement. They form aggregates of single, two, four and eight molecules. At high temperatures the one and two molecule aggregates dominate; as the temperature falls the larger clusters begin to dominate. The larger clusters occupy less space than the same number of molecules in smaller clusters. As a result, the density of water shows a maximum at 4°C.

Physical properties of most substances show uniform variation with temperature. In contrast, most physical properties of pure water show a minimum at some intermediate temperature. Sound velocity shows a maximum at 74°C

A list of some minimum temperatures

The physical property is given first, followed by the temperature in °C at which the minimum occurs:

oxygen solubility	80 °C
specific volume	4°C
specific heat	34°C
hydrogen solubility	37°C
compressibility	44°C
speed of light	-1°C
speed of sound (maximum)	74°C

When freezing, all water molecules form tetrahedrons. This leads to a sudden expansion in volume, ie a decrease in density. The solid phase of water is therefore lighter than the liquid phase, which is a rare property. consequences are:

The Concept of Salinity

Sea water contains 3.5% salts, dissolved gasses, organic substances and undissolved particulate matter. The presence of salts influences most physical properties of sea water like density, compressibility, freezing point, temperature of the density maximum to some degree but does not determine them. properties like viscosity, and light absorbtion) are not significantly affected by salinity. Particle and dissolved matter do affect light absorption in sea water and this influence is used in most optical applications. Two properties which are determined by the amount of salt in the sea are conductivity and osmotic pressure.

Ideally, salinity should be the sum of all dissolved salts in grams per kilogram of sea water. In practice, this is difficult to measure. The observation that - no matter how much salt is in the sea the various components present in a fixed ratio, helps to overcome the difficulty. It allows determination of salt content through the measurement of a substitution quantity and calculation of the total of all material making up the salinity from that measurement.

Determination of salinity could thus be made through its most important component, chloride. Chloride content was defined in 1902 as the total amount in grams of chlorine ions contained in one kilogram of sea water if all the halogens are replaced by chlorides. The definition reflects the chemical titration process for the determination of chloride content and is still of importance when dealing with historical data.

Salinity was defined in 1902 as the total amount in grams of dissolved substances contained in one kilogram of sea water if all carbonates are converted into oxides, all bromides and iodides into chlorides, and all organic substances oxidized. The relationship between salinity and chloride was determined through a series of fundamental laboratory measurements based on sea water samples from all regions of the world ocean and was given as:

$$S\ (^{\circ}/_{oo}) = 0.03 + 1.805\ Cl\ (^{\circ}/_{oo})$$

The symbol $^{\circ}/_{oo}$ stands for "parts per thousand" or "per ml"; a salt content of 3.5%^{0} is equivalent to 35 $^{\circ}/_{oo}$, or 35 grams of salt per kilogram of sea water.

The fact that the equation of 1902 gives a salinity of 0.03 $^{\circ}/_{oo}$ for zero chlorinity is a cause for concern. It indicates a problem in the water samples used for the laboratory measurements. The United Nations Scientific, Education and Cultural Organization (UNESCO) decided to repeat the base determination of the relation between chlorinity and salinity and introduced a new definition, known as absolute salinity ,

$$S\ (^{\circ}/_{oo}) = 1.80655\ Cl\ (^{\circ}/_{oo})\ (1969)$$

The definitions of 1902 and 1969 give identical results at a salinity of 35 $^{\circ}/_{oo}$ and do not differ significantly for most applications.

The definition of salinity was reviewed again when techniques to determine salinity from measurements of conductivity, temperature and pressure were developed. Since 1978, the "Practical Salinity Scale" defines salinity in terms of a conductivity ratio:

> " The practical salinity , symbol S, of a sample of sea water, is defined in terms of the ratio K of the electrical conductivity of a sea water sample of 15°C and the pressure of one standard atmosphere, to that of a potassium chloride (KCl)

solution, in which the mass fraction of KCl is 0.0324356, at the same temperature and pressure. The K value exactly equal to one corresponds, by definition, to a practical salinity equal to 35." The corresponding formula is:

$$S = 0.0080 - 0.1692\,K^{1/2} + 25.3853\,K + 14.0941\,K^{3/2} - 7.0261\,K^2 + 2.7081\,K^{5/2}$$

Note that in this definition, salinity is a ratio and ($^{\circ}/_{oo}$) is therefore no longer used, but an old value of $35^{\circ}/_{oo}$ corresponds to a value of 35 in the practical salinity. Some oceanographers cannot get used to numbers without units for salinity and write "35 psu", where psu is meant to stand for "practical salinity unit". As the practical salinity is a ratio and therefore does not have units, the unit "psu" is rather meaningless and strongly discouraged. Again, minute differences occur between the old definitions and the new Practical Salinity Scale, but they are usually negligible.

Electrical Conductivity of seawater

The conductivity of sea water depends on the number of dissolved ions per volume (i.e. salinity) and the mobility of the ions (ie temperature and pressure). Its units are mS/cm (milli-Siemens per centimetre). Conductivity increases by the same amount with a salinity increase of 0.01, a temperature increase of 0.01°C, and a depth (ie pressure) increase of 20 m. In most practical oceanographic applications the change of conductivity is dominated by temperature.

DENSITY

Density is one of the most important parameters in the study of the oceans' dynamics. Small horizontal density differences (caused for example by differences in surface heating) can produce very strong currents. The determination of density has therefore been one of the most important tasks in oceanography. The symbol for density is the Greek letter ρ (rho).

The density of sea water depends on temperature T, salinity S and pressure p. This dependence is known as the Equation of State of Sea Water.

The equation of state for an ideal gas was given by

$$p = \rho R T$$

where R is the gas constant. The exact equation for the entire range of temperatures, salinities and pressures encountered in the ocean

$$\tilde{n} = \rho(T,S,p)$$

(where S is salinity) is the result of many careful laboratory determinations. The first fundamental determinations to establish the equation were made in 1902 by Knudsen and Ekman. Their equation expressed ñ in g cm^{-3}. New fundamental determinations, based on data over a larger pressure and salinity range, resulted in a new density equation, known as the "International Equation of State". This equation uses temperature in °C, salinity from the Practical Salinity Scale and pressure in dbar (1 dbar = 10,000 pascal = 10,000 N m^{-2}) and gives density in kg m^{-3}. Thus, a density of 1.025 g cm^{-3} in the old formula corresponds to a density of 1025 kg m^{-3} in the International Equation of State.

Density increases with an increase in salinity and a decrease in temperature, except at temperatures below the density maximum. Oceanic density is usually close to 1025 kg m^{-3} (In freshwater it is close to 1000 kg m^{-3}). Oceanographers usually use the symbol $ó_t$ (the Greek letter sigma with a subscript t) for density, which they pronounce "sigma-t". This quantity is defined as $\sigma_t = \sigma - 1000$ and does not usually carry units (it should carry the same units as ñ). A typical seawater density is thus $\sigma_t = 25$. A useful rule of thumb is that σ_t changes by the same amount if T changes by 1°C, S by 0.1, and p by the equivalent of a 50 m depth change.

Notice that the density maximum is above the freezing point for salinities below 24.7 but below the freezing point for salinities above 24.7. This affects the thermal convection:

- S < 24.7: The water cools until it reaches maximum density; then, when the surface water becomes lighter (ie after the density maximum has been passed) cooling is restricted to the wind-mixed layer, which eventually freezes over. The deep basins are filled with water of maximum density.
- S > 24.7: Convection always reaches the entire water body. Cooling is slowed down because a large amount of heat is stored in the water body. This is because the water reaches freezing point before the maximum density is attained.

If your browser supports JavaScript you can check the range of seawater density and its dependence on temperature and salinity at surface pressure with this density calculator: Enter a value for temperature, a value for salinity and press the calculate button. Verify your result against the appropriate TS-diagram.

Physical properties

Water has several important physical properties. Although these properties are familiar because of the omnipresence of water, most of the physical properties of water are quite atypical. Given the low molar mass of its constituent molecules, water has unusually large values of viscosity, surface tension, heat of vaporization, and entropy of vaporization, all of which can be ascribed to the extensive hydrogen bonding interactions present in liquid water. The open structure of ice that allows for maximum hydrogen bonding explains why solid water is less dense than liquid water—a highly unusual situation among common substances.

Selected physical properties of water

molar mass	18.0151 grams per mole
melting point	0.00 °C
boiling point	100.00 °C
maximum density (at 3.98 °C)	1.0000 grams per cubic centimetre
density (25 °C)	0.99701 grams per cubic centimetre
vapour pressure (25 °C)	23.75 torr
heat of fusion (0 °C)	6.010 kilojoules per mole
heat of vaporization (100 °C)	40.65 kilojoules per mole
heat of formation (25 °C)	-285.85 kilojoules per mole
entropy of vaporization (25 °C)	118.8 joules per °C mole
viscosity	0.8903 centipoise
surface tension (25 °C)	71.97 dynes per centimeter

WHY IS SEAWATER SALTY?

Salt in the ocean comes from two sources: runoff from the land and openings in the seafloor.

Rocks on land are the major source of salts dissolved in seawater. Rainwater that falls on land is slightly acidic, so it erodes rocks. This releases ions that are carried away to streams and rivers that eventually feed into the ocean. Many of the dissolved ions are used by organisms in the ocean and are removed from the water. Others are not removed, so their concentrations increase over time.

Another source of salts in the ocean is hydrothermal fluids, which come from vents in the seafloor. Ocean water seeps into cracks in the seafloor and is heated by magma from the Earth's core. The heat causes a series of chemical reactions. The water tends to lose oxygen, magnesium, and sulfates, and pick up metals such as iron, zinc, and copper from surrounding rocks. The heated water is released through vents in the seafloor, carrying the metals with it. Some ocean salts come from underwater volcanic eruptions, which directly release minerals into the ocean.

Salt domes also contribute to the ocean's saltiness. These domes, vast deposits of salt that form over geological timescales, are found underground and undersea around the world. They are common across the continental shelf of the northwestern Gulf of Mexico.

Two of the most prevalent ions in seawater are chloride and sodium. Together, they make up around 85 percent of all dissolved ions in the ocean. Magnesium and sulfate make up another 10 percent of the total. Other ions are found in very small concentrations. The concentration of salt in seawater (salinity) varies with temperature, evaporation, and precipitation. Salinity is generally low at the equator and at the poles, and high at mid-latitudes. The average salinity is about 35 parts per thousand. Stated in another way, about 3.5 percent of the weight of seawater comes from the dissolved salts.

WHY DOES SEAWATER SALINITY VARY?

Salinity distribution

A discussion of salinity, the salt content of the oceans, requires an understanding of two important concepts: (1) the present-day oceans are considered to be in a steady state, receiving as much salt as they lose, and (2) the oceans have been mixed over such a long time period that the composition of sea salt is the same everywhere in the open ocean. This uniformity of salt content results in oceans in which the salinity varies little over space or time.

The range of salinity observed in the open ocean is from 33 to 37 grams of salt per kilogram of seawater or psu. For the most part, the observed departure from a mean value of approximately 35 psu is caused by processes at Earth's surface that locally add or remove fresh water. Regions of high evaporation have elevated surface salinities, while regions of high precipitation have depressed surface salinities. In nearshore regions close to large freshwater sources, the salinity may be lowered by dilution. This is especially true in areas where the region of the ocean receiving the fresh water is isolated from the open ocean by the geography of the land.

Areas of the Baltic Sea may have salinity values depressed to 10 psu or less. Increased salinity by evaporation is accentuated where isolation of the water occurs. This effect is found in the Red Sea, where the surface salinity rises to 41 psu. Coastal lagoon salinities in areas of high evaporation may be much higher. The removal of fresh water by evaporation or the addition of fresh water by precipitation does not affect the constancy of composition of the sea salt in the open sea. A river draining a particular soil type, however, may bring to the oceans only certain salts that will locally alter the salt composition. In areas of high evaporation where the salinity is driven to very high values, precipitation of particular salts may alter the composition too. At high latitudes where sea ice forms seasonally and icebergs are often released into the open ocean, the salinity of the seawater is reduced when ice melts and is elevated during ice formation. This saltier water can then sink down into the deep ocean (*see* density current).

Salinity Patterns

All of the salts and ions that dissolve in seawater contribute to its overall salinity. Salinity of seawater is usually expressed as the grams of salt per kilogram (1000 g) of seawater. On average, about 35 g of salt is present in each 1 kg of seawater, so we say that the average salinity of the ocean salinity is 35 parts per thousand (ppt). Note that 35 ppt is equivalent to 3.5% (parts per hundred). Some sources now use practical salinity units (PSU) to express salinity values, where 1 PSU = 1 ppt. The units are not included, so we can refer simply to a salinity of 35.

Many different substances are dissolved in the ocean, but six ions comprise about 99.4% of all the dissolved ions in seawater. These six **major ions** are (Table 3):

Table 3. The six major ions in seawater

	g/kg in seawater	% of ions by weight
Chloride Cl^-	19.35	55.07%
Sodium Na^+	10.76	30.6%
Sulfate SO_4^{2-}	2.71	7.72%
Magnesium Mg^{2+}	1.29	3.68%
Calcium Ca^{2+}	0.41	1.17%
Potassium K^+	0.39	1.1%
		99.36%

Chloride and sodium, the components of table salt (sodium chloride NaCl), make up over 85% of the ions in the ocean, which is why seawater tastes salty (Fig. 10). In addition to the major constituents, there are numerous minor constituents; radionucleotides, organic compounds, metals etc. These minor constituents are found in concentrations of ppm (parts per million) or ppb (parts per billion), unlike the major ions that are far more abundant (ppt) (Table 4). To put this into perspective, 1 ppm = 1 mg/kg, or the equivalent of 1 teaspoon of sugar dissolved in 14,000 cans of soda. 1 ppb = 1 ìg/kg, or the equivalent of 1 teaspoon of a substance dissolved in five Olympic-sized swimming pools! These minor constituents represent numerous substances, but together they make up less than 1% of the ions in the seawater. Some of these may be important as minerals and trace elements vital to living organisms, but they don't have much impact on the overall salinity. But given the vast size of the oceans, even materials found in trace abundance can represent fairly large reservoirs. For example gold is a trace element in seawater, found in concentrations of parts per trillion, yet if you could extract all of the gold in just one km^3 of seawater, it would be worth about $20 million.

Table 4 Concentrations of some minor elements in seawater

	g/kg in seawater		g/kg in seawater
Carbon	0.028	Iron	2 x 10-6
Nitrogen	0.0115	Manganese	2 x 10-7
Oxygen	0.006	Copper	1 x 10-7
Silicon	0.002	Mercury	3 x 10-8

	g/kg in seawater		g/kg in seawater
Phosphorous	6 x 10-5	Gold	4 x 10-9
Uranium	3.2 x 10-6	Lead	5 x 10-10
Aluminum	2 x 10-6	Radon	6 x 10-19

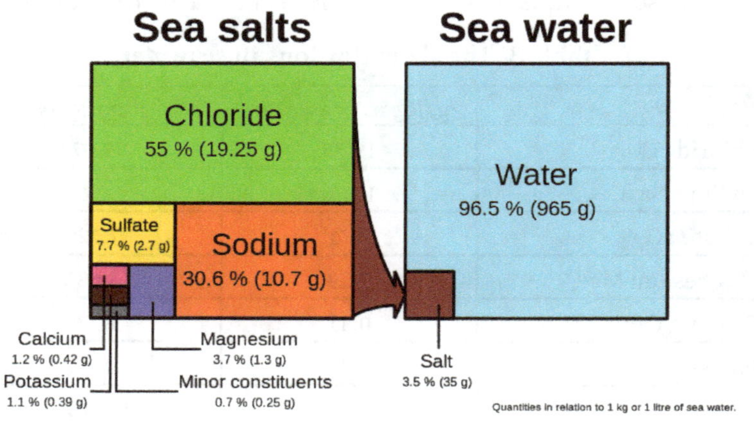

Fig. 10. The relative proportions of ions in seawater.

Because the six major ions in seawater comprise over 99% of the total salinity, changes in abundance of the minor constituents have little effect on overall salinity. Furthermore, the **rule of const**ant proportions states that even though the absolute salinity of ocean water might differ in different places, the relative proportions of the six major ions within that water are always constant. For example, no matter the total salinity of a seawater sample, 55% of the total salinity will be due to chloride, 30% due to sodium, and so on. Since the proportion of these major ions does not change, we call these conservative ions.

Given these constant proportions, in order to calculate total salinity you can simply measure the concentration of just one of the major ions and use that value to calculate the rest. Traditionally chloride has been the ion measured because it is the most abundant, and thus the simplest to measure accurately. Multiplying the concentration of chloride by 1.8 gives the total salinity. For example, looking at Fig. 10, 19.25 g/kg (ppt) chloride x 1.8 = 35 ppt. Today, for rapid measurements of salinity, electrical conductivity is often used rather than determining chloride concentrations

It is important to be aware that while the rule of constant proportions applies to most of the ocean, there may be certain coastal areas where lots of river discharge may alter these proportions slightly. Furthermore, it is important to remember that the rule of constant proportions only applies to the major ions. The proportions of the minor ions may fluctuate, but remember that they make a very minor contribution to overall salinity. Because the concentrations of the minor ions are not constant, these are referred to as non-conservative ions.

Why are the major ions found in constant proportions? There is constant input of ions from river runoff and other processes, usually in very different proportions from what is found in the ocean. So why don't the proportions in the oceans change? Most of the ions discharged by rivers have fairly low residence times compared to ions in seawater, usually because they are used in biological processes. These low residence times do not allow the ions to accumulate and alter salinity. Also, the mixing time of the world ocean is around 1000 years, which is very short compared to the residence times of the major ions, which may be tens of millions of years long. So during the residence time of a single ion the ocean has mixed numerous times, and the major ions have become evenly distributed throughout the ocean.

HOW DOES SEAWATER SALINITY VARY AT THE SURFACE AND WITH DEPTH?

Factors Affecting Ocean Salinity

- The salinity of water in the surface layer of oceans depend mainly on evaporation and precipitation.
- Surface salinity is greatly influenced in coastal regions by the fresh water flow from rivers, and in polar regions by the processes of freezing and thawing of ice.
- Wind, also influences salinity of an area by transferring water to other areas.
- The ocean currents contribute to the salinity variations.
- Salinity, temperature and density of water are interrelated. Hence, any change in the temperature or density influences the salinity of an area.

Total salinity in the open ocean averages 33-37 ppt, but it can vary significantly in different locations. But since the major ion proportions are constant, the regional salinity differences must be due more to water input and removal rather than the addition or removal of ions. Fresh water input comes through processes like precipitation, runoff from land, and melting ice. Fresh water removal primarily comes from evaporation and freezing (when seawater freezes, the resulting ice is mostly fresh water and the salts are excluded, making the remaining water even saltier). So differences in rates of precipitation, evaporation, river discharge, and ice formation play a significant role in regional salinity variations. For example, the Baltic Sea has a very low surface salinity of around 10 ppt, because it is a mostly enclosed body of water with lots of river input. Conversely, the Red Sea is very salty (around 40 ppt), due to the lack of precipitation and the hot environment which leads to high levels of evaporation.

One of the saltiest large bodies of water on Earth is the Dead Sea, between Israel and Jordan. Salinity in the Dead Sea is around 330 ppt, which is almost ten times saltier than the ocean. This extremely high salinity is a result of the hot, arid conditions in the Middle East that lead to high rates of evaporation. In addition, in the 1950s the flow from the Jordan River was diverted away from the Dead Sea, so there is no longer significant fresh water input. With no input and high evaporation, the water level in the Dead Sea is receding at a rate of about 1 m per year. The high salinity makes the water very dense, which creates buoyant forces that allow people to easily float at the surface. But the high salinity also means that the water is too salty for most living organisms, so only microbes are able to call it home; hence the name the Dead Sea. But as salty as the Dead Sea may be, it is not the saltiest body of water on Earth. That distinction currently belongs to Gaet'ale Pond in Ethiopia, with a salinity of 433 ppt.

Latitudinal Variations

While local conditions are important for determining salinity patterns in any single location, there are some global patterns that bear further investigation. Temperature is highest at the equator, and lowest near the poles, so we would expect higher rates of evaporation, and therefore higher salinity, in equatorial regions. This is generally the case, but in the figure below salinity right along the equator seems to be a little lower than at slightly higher latitudes. This is because equatorial regions also get a high volume of rain on a regular basis, which dilutes the surface water along the equator. So the higher salinities are found at subtropical, warm latitudes with high evaporation and less precipitation. At the poles there is little evaporation, which, coupled with ice and snow melting, produces a relatively low surface salinity. The image below shows high salinity in the Mediterranean Sea; this is located in a warm region with high evaporation, and the sea is largely isolated from mixing with the rest of the North Atlantic water, leading to high salinity. Lower salinities, such as those around southeast Asia, are the result of precipitation and high volumes of river input.

Vertical Variation

In addition to geographical variation in salinity, there are also changes in salinity with depth. As we have seen, most differences in salinity are due to variations in evaporation, precipitation, runoff, and ice cover. All of these process occur at the ocean surface, not at depth, so the most pronounced differences in salinity should be found in surface waters. Salinity in deeper water remains relatively uniform, as it is unaffected by these surface processes. Some representative salinity profiles are shown in Fig. 11. At the surface, the top 200 m or so show relatively uniform salinity in what is called the **mixed layer**. Winds, waves, and surface currents stir up the surface water, causing a great deal of mixing in this layer and fairly uniform salinity conditions. Below the mixed layer is an area of rapid salinity change over a small change in depth. This zone of rapid change is called the **halocline**, and it represents a transition between the mixed layer and the deep ocean. Below the halocline, salinity may show little variation down to the seafloor, as this region is far removed from the surface processes that impact salinity. In the figure below, note the low surface salinity at high latitudes, and higher surface salinity at low latitudes as discussed above. Yet despite the surface differences, salinity at depth in both locations may be very similar.

Vertical Distribution of Salinity

- Salinity changes with depth, but the way it changes depends upon the location of the sea.

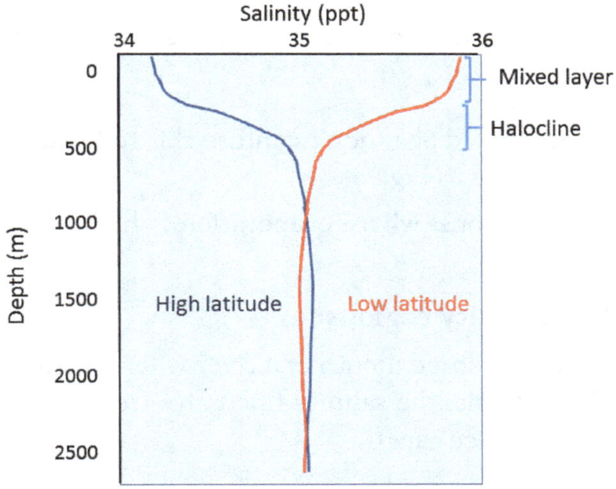

Fig. 11. Salinity profiles from two hypothetical sites in the open ocean, one from high latitude and one from low latitude

- Salinity at the surface increases by the loss of water to ice or evaporation, or decreased by the input of fresh waters, such as from the rivers.
- Salinity at depth is very much fixed, because there is no way that water is 'lost', or the salt is 'added.' There is a marked difference in the salinity between the surface zones and the deep zones of the oceans.
- The lower salinity water rests above the higher salinity dense water.
- Salinity, generally, increases with depth and there is a distinct zone called the **halocline** (compare this with thermocline), where salinity increases sharply.
- Other factors being constant, increasing salinity of seawater causes its density to increase. High salinity seawater, generally, sinks below the lower salinity water. This leads to **stratification by salinity**.

Sub-Surface Salinity

- With depth, the salinity also varies, but this variation again is subject to latitudinal difference. The decrease is also influenced by cold and warm currents.
- In high latitudes, salinity increases with depth. In the middle latitudes, it increases up to 35 metres and then it decreases. At the equator, surface salinity is lower.

Horizontal distribution of salinity

To make life easier, remove the symbol $^{o}/_{oo}$ and place only number

- The salinity for normal open ocean ranges between 33 and 37.

High salinity regions

- In the land locked Red Sea (don't confuse this to Dead Sea which has much greater salinity), it is as high as 41.
- In hot and dry regions, where evaporation is high, the salinity sometimes reaches to 70.

Comparatively Low salinity regions

- In the estuaries (enclosed mouth of a river where fresh and saline water get mixed) and the Arctic, the salinity fluctuates from 0 – 35, seasonally (fresh water coming from ice caps).

Pacific

- The salinity variation in the Pacific Ocean is mainly due to its shape and larger areal extent.

Atlantic

- The average salinity of the Atlantic Ocean is around 36-37.
- The equatorial region of the Atlantic Ocean has a salinity of about 35.
- Near the equator, there is heavy rainfall, high relative humidity, cloudiness and calm air of the doldrums.
- The polar areas experience very little evaporation and receive large amounts of fresh water from the melting of ice. This leads to low levels of salinity, ranging between 20 and 32.
- Maximum salinity is observed between 20° N and 30° N and 20° W – 60° W. It gradually decreases towards the north.

HOW DOES SEAWATER DENSITY VARY WITH DEPTH?

The density of a material is given in units of mass per unit volume and expressed in kilograms per cubic metre in the SI system of units. In oceanography the density of seawater has been expressed historically in grams per cubic centimetre. The density of seawater is a function of temperature, salinity, and pressure. Because oceanographers require density measurements to be accurate to the fifth decimal place, manipulation of the data requires writing many numbers to record each measurement. Also, the pressure effect can be neglected in many instances by using potential temperature. These two factors led oceanographers to adopt a density unit called sigma-*t* (σ_t). This value is obtained by subtracting 1.0 from the density and multiplying the remainder by 1,000. The $\acute{o}_t$ has no units and is an abbreviated density of seawater controlled by salinity and temperature only. The σ_t of seawater increases with increasing salinity and decreasing temperature.

The relationship between pressure and density is demonstrated by observing the effect of pressure on the density of seawater at 35 psu and 0 °C. Because a one-metre (three-foot) column of seawater produces a pressure of about one decibar (0.1 atmosphere), the pressure in decibars is approximately equal to the depth in metres. (One decibar is one-tenth of a bar, which in turn is equal to 10^5 newtons per square metre.)

Values associated with the change in seawater density with depth are listed in Table 5.

Table 5. Density changes with depth (seawater 35 parts per thousand and 0 °C)

depth (m)	pressure (decibars)	density (g/cm3)
0	0	1.02813
1,000	1,000	1.03285
2,000	2,000	1.03747
4,000	4,000	1.04640
6,000	6,000	1.05495
8,000	8,000	1.06315
10,000	10,000	1.07104

Increasing density values demonstrate the compressibility of seawater under the tremendous pressures present in the deep ocean. If seawater were incompressible, each cubic centimetre of water in the water column would expand, and density values at all depths would be equal. If the average pressure occurring at a depth of 4,000 metres (about 13,100 feet, the approximate mean depth of the ocean) was somehow replaced with the average pressure that occurred at 2,000 metres (about 6,600 feet) and the area of the oceans remained constant, there would be an average sea level rise of about 36 metres (120 feet).

VARIATION IN DENSITY

Seawater density varies from place to place because it is affected by salinity and temperature. This means that ships float higher or lower in the water, depending on the density of the ocean. If you look near the waterline of a cargo ship, you should find the International Load Line, once called the Plimsoll Line. This shows the limit of where the fully loaded ship should sit in waters of different densities.

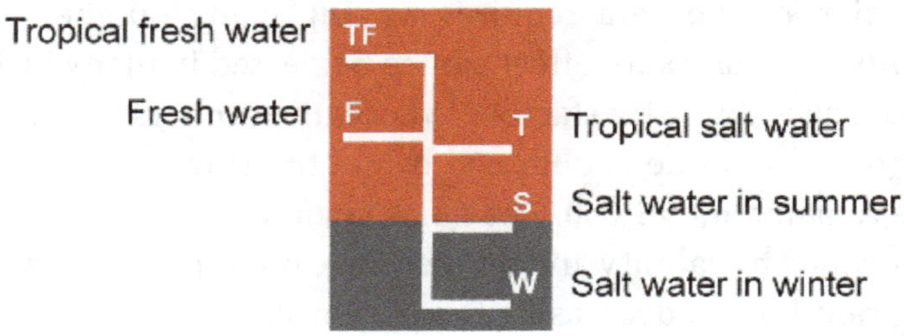

Fig. 12. Ship's load line

High salinity makes water denser. This is because there is more salt packed into the water.

High temperature makes water less dense. As water gets warmer, its molecules spread out, so it becomes less dense. As it gets colder, it becomes denser. Most chemicals get denser when they turn from a liquid to a solid, but water is different. When liquid water freezes into solid ice, it becomes less dense. When ice forms, water molec structure is less dense th

Deep wa are packed together mo n.

What happ

Dense wa t drives the deep ocean c high salinity and low tem ink into the deep ocean a

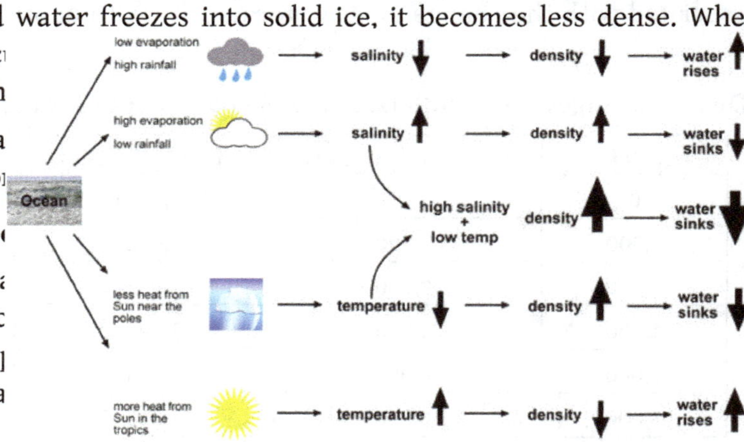

Fig. 13. Ocean salinity, temperature and density

IS THE OCEAN ACIDIC OR BASIC?

Until recently, the amount of carbon dioxide in the atmosphere has fluctuated slightly and slowly during the past 10,000 years. However, the Industrial Revolution of the 1700s started a global adoption of fossil fuels to power human activity. The rate at which fossil fuels like coal, oil, and natural gas are burned has increased up until the present day.

Burning fossil fuels releases carbon dioxide gas to the atmosphere, and the ever-increasing global use of fossil fuels has caused the amount of carbon dioxide in the atmosphere to increase to a concentration that is higher than any time in the past 800,000 years.

The cutting of forests for fuel or to clear land for agriculture over the past 250 years has also contributed to higher carbon dioxide levels in the atmosphere because trees capture and store carbon dioxide via photosynthesis.

Not only do higher atmospheric concentrations of carbon dioxide alter the Earth's climate, they also impact ocean chemistry. This is because carbon dioxide in the atmosphere readily dissolves into water.

Dissolved Carbon Dioxide: Gases in Liquid?

Just as solids like sugar can dissolve in water, gases like carbon dioxide do as well. This idea is easily demonstrated in a bottle of soda. The manufacturer dissolves carbon dioxide in the beverage. The dissolved carbon dioxide is invisible to the naked eye, but once the bottle is opened carbon dioxide escapes as bubbles that tickle your nose. The extra carbon dioxide in soda water adds acidity to the liquid. Similarly, about one third of the carbon dioxide gas in Earth's atmosphere dissolves into the oceans.

Carbon Dioxide Imparts Acidity: Transformations of Carbon Dioxide in Water

Once carbon dioxide dissolves in water, it reacts with water molecules to form carbonic acid. Carbonic acid can be further transformed to bicarbonate and carbonate ions. These four different forms of carbon (dissolved carbon dioxide, carbonic acid, bicarbonate, and carbonate) exist in balanced proportions in seawater.

As more carbon dioxide is added to seawater, the balance shifts and the carbonate ion concentration decreases as it is transformed to bicarbonate due to increasing acidity.

How Acidity is Measured: pH

The acidity of a liquid is reported as pH. The lower the pH value, the higher the acidity of a liquid. Solutions with low pH are acidic and solutions with high pH are basic (also known as alkaline).

Prior to the Industrial Revolution, average ocean pH was about 8.2. Today, average ocean pH is about 8.1. This might not seem like much of a difference, but the relationship between pH and acidity is not direct. Each decrease of one pH unit is a ten-fold increase in acidity. This means that the acidity of the ocean today, on average, is about 25% greater than it was during preindustrial times.

Acidity and Availability of Shell-forming Calcium

Carbonate

Marine life uses carbonate from the water to build shells and skeletons. As seawater becomes more acidic, carbonate is less available for animals to build shells and skeletons. Under conditions of severe acidification, shells and skeletons can dissolve.

Coastal Acidification

Closer to Home: Coastal Acidification

Human activity also contributes to acidification in coastal waters. Acid-forming compounds (including carbon dioxide) are released to the atmosphere when fossil fuels are burned, and excess nutrients contribute to acidification in coastal waters when algal blooms peak and die.

Acid Rain

Burning fossil fuels for energy releases water and carbon dioxide as the main byproducts, but nitrogen oxides and sulfur dioxide are also released in smaller amounts. These two acid-forming compounds fall back to Earth's surface. They can land in coastal waters directly, or more often mix with water in the atmosphere before falling as acid rain. Acid rain typically has a pH between 4.2 and 4.4.

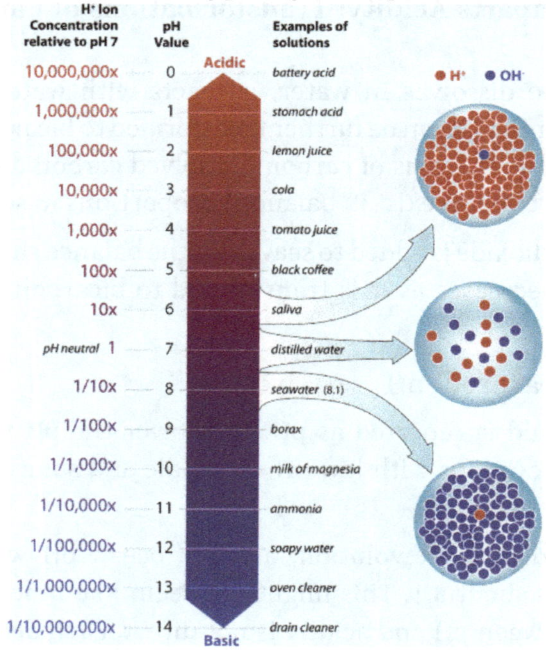

Fig. 14. The acidity of a liquid results from its concentration of hydrogen ions, which is typically measured and reported as pH.

Excess Nutrients Delivered Via Streams

The elements nitrogen and phosphorous are essential nutrients for living things. For this reason ,farmers homeowners, and gardeners supply nitrogen and phosphorous to crops, lawns, and gardens to stimulate plant growth. However, water can carry excess nutrients down streams and into coastal waters.

Agricultural activities are a major source of nutrients to coastal waters, but other sources include sewage, wastewater treatment plant effluent, and nitrogen oxide air pollution. In coastal waters excess nutrients stimulate the growth of algae. Algae multiply rapidly under ideal growing conditions and algal blooms can impair water quality by causing hypoxia, foul odors and even toxins.

A less well-known fact is that algal blooms can contribute to acidification. When algae die their decomposing tissue releases carbon dioxide directly into the water, resulting in acidification.

WHAT FACTORS CONTROL THE DISTRIBUTION OF CARBON AND OXYGEN IN THE OCEAN?

The idea seemed simple enough: the more carbon dioxide that people pumped into the atmosphere by burning fossil fuels, the more the oceans would absorb. The ocean would continue to soak up more and more carbon dioxide until global warming heated the ocean enough to slow down ocean circulation. Water trapped at the surface would become saturated, at which point, the ocean would slow its carbon uptake. To oceanographers of 30 years ago, the question was less, how will human emissions change the ocean carbon cycle, and more, is the ocean carbon cycle changing yet?

The question matters because if the ocean starts to take up less carbon because of global warming, more is left in the atmosphere where it can contribute to additional warming. Scientists wanted to understand how the ocean carbon cycle might change so that they could make more accurate predictions about global warming. Thus motivated, oceanographers began a series of research cruises, trolling across the Pacific from Japan to California, from Alaska to Hawaii, and through the North Atlantic from Europe to North America. On shore, others developed computer models.

After 30 years of research, the question itself hasn't changed, but the reasoning behind it couldn't be more different. Oceanographers started out wanting to know if the ocean was keeping up with the amount of carbon dioxide people are putting into the atmosphere. Instead, they found that people aren't the only players changing the ocean carbon cycle.

Over decades, natural cycles in weather and ocean currents alter the rate at which the ocean soaks up and vents carbon dioxide. What's more, scientists are beginning to find evidence that human-induced changes in the atmosphere also change the rate at which the ocean takes up carbon. In other words, it turns out that the world is not a simple place.

The Measured Ocean

The group surrounds a circular cluster of instruments and 36 three-foot-tall PVC (plastic) bottles, taking turns extracting sea water from the bottles, assembly-line style. It is a deliberate, well-ordered procedure. The glass sample bottles set aside for oxygen samples are filled first, followed by the massive syringe meant for chlorofluorocarbon (freon) samples, and so on, until 10 to 15 different samples have come out of each bottle. Everyone has a task and a place. It's a social event, a break from the lonely hours each will spend in his or her lab analyzing the samples before the next batch is hauled out of the ocean. It might even be fun. Except that it's late winter. In the North Pacific. And they are on the deck of a ship, looking at the same faces that they've seen day after day for four weeks or more, and they'll be repeating this procedure again in another 30 nautical miles.

"It's pretty brutal," says Richard Feely, with the air of a veteran who thoroughly enjoys his work. An oceanographer who studies the ocean carbon cycle at the National Oceanic and Atmospheric Administration's Pacific Marine Environmental Laboratory, he's been at this sort of thing for almost forty years. Feely is one of a community of oceanographers who have been monitoring Earth's oceans for decades, trying to figure out how much human-released carbon dioxide the ocean has been soaking up.

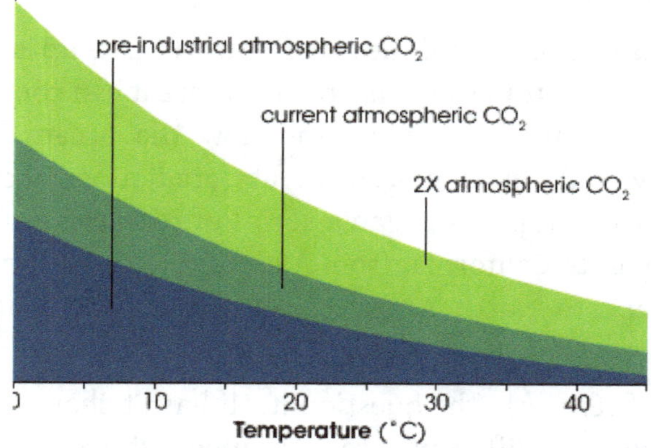

Fig. 15. The concentration of carbon dioxide (CO_2) in ocean water (y axis) depends on the amount of CO_2 in the atmosphere (shaded curves) and the temperature of the water (x axis). This simplified graph shows that as atmospheric CO_2 increases from pre-industrial levels (blue) to double (2X) the pre industrial amounts (light green), the ocean CO_2 concentration increases as well. However, as water temperature increases, its ability dissolve CO_2 decreases.

For eons, the world's oceans have been sucking carbon dioxide out of the atmosphere and releasing it again in a steady inhale and exhale. The ocean takes up carbon dioxide through photosynthesis by plant-like organisms (phytoplankton), as well as by simple chemistry: carbon dioxide dissolves in water. It reacts with seawater, creating carbonic acid. Carbonic acid releases hydrogen ions, which combine with carbonate in seawater to form bicarbonate, a form of carbon that doesn't escape the ocean easily.

As we burn fossil fuels and atmospheric carbon dioxide levels go up, the ocean absorbs more carbon dioxide to stay in balance. But this absorption has a price: these reactions lower the water's pH, meaning it's more acidic. And the ocean has its limits. As temperatures rise, carbon dioxide leaks out of the ocean like a glass of root beer going flat on a warm day. Carbonate gets used up and has to be re-stocked by upwelling of deeper waters, which are rich in carbonate dissolved from limestone and other rocks.

In the center of the ocean, wind-driven currents bring cool waters and fresh carbonate to the surface. The new water takes up yet more carbon to match the atmosphere, while the old water carries the carbon it has captured into the ocean.

The warmer the surface water becomes, the harder it is for winds to mix the surface layers with the deeper layers. The ocean settles into layers, or stratifies. Without an infusion of fresh carbonate-rich water from below, the surface water saturates with carbon dioxide. The stagnant water also supports fewer phytoplankton, and carbon dioxide uptake from photosynthesis slows. In short, stratification cuts down the amount of carbon the ocean can take up.

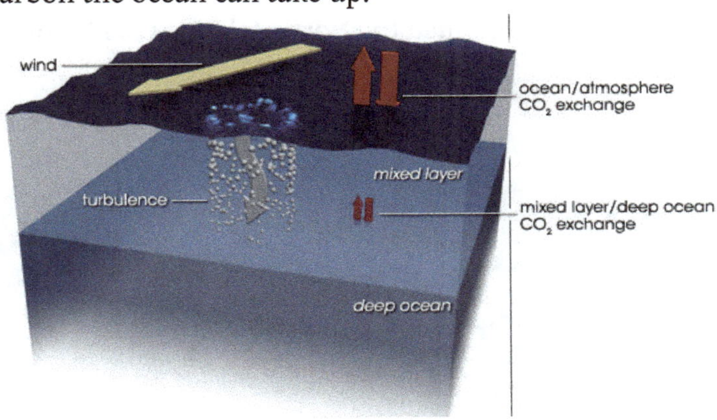

Fig. 16. In the short term, the ocean absorbs atmospheric carbon dioxide into the mixed layer, a thin layer of water with nearly uniform temperature, salinity, and dissolved gases. Wind-driven turbulence maintains the mixed layer by stirring the water near the ocean's surface. Over the long term, carbon dioxide slowly enters the deep ocean at the bottom of the mixed layer as well in in regions near the poles where cold, salty water sinks to the ocean depths.

Defying Expectations: To try to understand the ocean's carbon limits, Feely and the rest of the oceanography community measure dissolved carbon and the ocean's pH. They probe its temperature, alkalinity, salinity, and record the presence of tracers like chlorofluorocarbons or helium to find out when the water was last exposed to the atmosphere. By the end of a typical cruise, they will have collected 50,000 or more measurements. And then they go out the next year to repeat it all again, and they have done this for more than three decades.

"When we started in the 70s and 80s, we had this concept: we'll measure [carbon dioxide concentrations in the ocean] 10 years later, and we'll just see the anthropogenic input," says Feely. "We had very simplistic ideas that the anthropogenic changes would be the only changes we would see," he adds a little ruefully.

Feely and his colleagues saw changes, but they weren't at all the changes they expected. Carbon concentrations in the ocean did rise as atmospheric carbon dioxide skyrocketed, but in 2006, Feely and several colleagues announced that the equatorial Pacific seemed to be venting more carbon dioxide to the atmosphere between 1997 and 2004 than it had in previous years. And in 2007, Ute Schuster and Andrew Watson, oceanographers from the University of East Anglia, reported that amount of carbon that the North Atlantic Ocean soaked up decreased by a factor of two between 1994 and 2005. The ocean, or parts of it, seemed to be taking up less, not more, carbon.

Had the ocean already stratified, slowing the rate at which it soaked up carbon? Schuster and Watson believed they saw stratification at work in the North Atlantic, but the drop in the amount of carbon being taken up was too large for global warming to be acting alone. During the decade Schuster and Watson made their observations, a large-scale weather pattern, called the North Atlantic Oscillation, shifted.

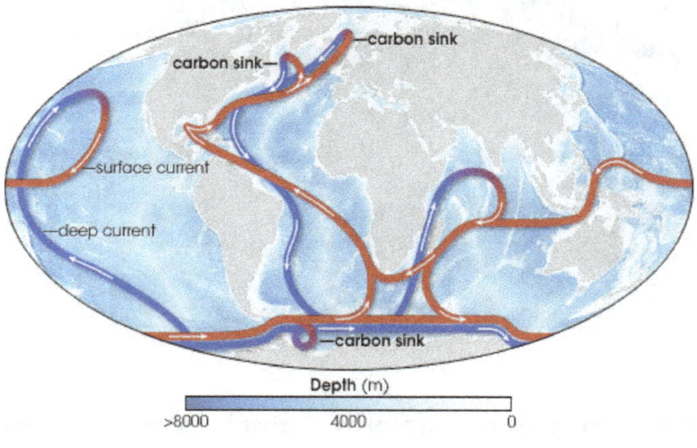

Fig. 17. The global oceans are connected by deep currents (blue lines) and surface currents (red). Carbon from the atmosphere enters the ocean depths in areas of deep water formation in the North Atlantic and offshore of the Antarctic Peninsula. Where deep currents rise towards the surface, they can release "fossil" carbon dioxide stored centuries ago.

Like El Niño in the Pacific, the North Atlantic Oscillation (NAO) changes weather on a large scale. In the early 1990s, says Watson, the North Atlantic Oscillation brought stronger and more frequent winds to the northern regions of the North Atlantic during the winter. The winds stirred the ocean, tucking carbon-dioxide-laden surface water down and pulling unsaturated water to the surface, in effect increasing the rate at which the North Atlantic took up carbon. By 2000, the North Atlantic Oscillation shifted, calming winds and allowing warmer waters to expand north. These two changes increased stratification in the North Atlantic and slowed the carbon uptake between 1994 and 2005.

The ocean does not take up carbon uniformly. It breathes, inhaling and exhaling carbon dioxide. In addition to the wind-driven currents that gently stir the center of ocean basins (the waters that are most limited by stratification), the ocean's natural, large-scale circulation drags deep water to the surface here and there. Having collected carbon over hundreds of years, this deep upwelling water vents carbon dioxide to the atmosphere like smoke escaping through a chimney. The stronger upwelling brought by the cold phase of the Pacific Decadal Oscillation apparently enhanced the size of the chimney and let more carbon escape to the atmosphere.

Control of distribution of dissolved organic carbon in the surface ocean

Around 662 billion tons of organic carbon are dissolved in the ocean, making the pool one of Earth's major, exchangeable carbon reservoirs. Dissolved organic carbon (DOC) has many ecological functions. It can form complexes with metals; absorb UV and visible light, acting as a "sunscreen" for marine microorganisms and controlling primary production in the upper water column; it has antioxidant activity, reacting with free radicals in the media; but most importantly, it serves as substrate for the microbial loop and as a vehicle for carbon sequestration in the ocean. Therefore, DOC plays an important role in climate on geological time scales.

Because the amount of atmospheric CO_2 is of the same magnitude as the DOC pool, and is closely linked to it through exchange, variations in one of these reservoirs can affect the other, impacting the carbon cycle with consequences for climate. Significant net DOC remineralization would lead to an increase of atmospheric CO_2, enhancing greenhouse warming at the surface of the Earth. Net oxidation of only 1% of the seawater DOC pool within 1 year would be sufficient to generate a CO_2 flux of 7 PgC/yr, comparable to that produced annually by fossil fuel combustion. It has also been proposed that a large-scale oxidation of DOC may have prevented a dramatic global glaciation ('snowball earth') in the Neoproterozoic period.

Despite its importance, knowledge about DOC dynamics is relatively limited; in fact, it was considered highly inert until about three decades ago when a new analytical technique for measuring it via high-temperature catalytic oxidation stimulated new interest. The technique eventually provided more accurate DOC values, showing that it was more involved in the carbon cycle than previously thought and that its concentrations vary with depth, time, and location. Considering DOC distributions observed in the surface Atlantic Ocean, we see values in the subtropical gyres of 65-70 μmol kg^{-1}, the highest concentrations in the tropics (> 70 μmol kg^{-1}), the lowest in the Southern Ocean (< 50 μmol kg^{-1}), and moderate concentrations in the northern North Atlantic (55-60 μmol kg^{-1}); this pattern is consistent in other ocean basins. So what controls this distribution and can we predict it? Even with improved analytical techniques, DOC is not a variable that can be measured easily at sea, and the sampling must be done carefully since it is easy to contaminate. Therefore, DOC data are typically fewer than those of other more readily determined variables such as nutrients and oxygen. If we could predict DOC from variables for which much greater global ocean coverage exists, we could fill the very large spatial and temporal gaps in the DOC fields.

DOC is produced in the upper water column by phytoplankton (primary producers). Actually, half of the inorganic carbon that is fixed by phytoplankton is transformed to DOC. Heterotrophic microbes consume most of that DOC, but ~ 4% of global annual net primary production (~ 2 Pg C y^{-1}) accumulates as DOC, much of which is exported to the mesopelagic via vertical mixing and convergence, thus contributing to the biological carbon pump.

New primary production, the foundation of a system's net community production (NCP), depends on new nutrients reaching the euphotic zone, which happens primarily via upwelling in divergence zones and winter vertical mixing. NCP is the balance of the carbon generated by primary producers minus that lost through heterotrophic respiration (prokaryotes and animals). It can be estimated either by a loss of reactants (CO_2 or nutrients) or a gain in products (suspended POC, DOC, and export production).

In our work, we needed to establish the fraction of NCP that was present in dissolved form (i.e., the net DOC production ratio, or NDPr). For that, we simply estimated NCP from the nitrate (NO_3^-) that is consumed in the euphotic zone (DNO_3^-):

$$\Delta NO_3^- = \text{new } NO_3^- \text{ (introduced from deeper layers)} - \text{remaining } NO_3^- \text{ (at surface)} \quad \text{(Eq. 1)}$$

In the same way, we also calculated net accumulated DOC, or ΔDOC:

$$\Delta DOC = \text{DOC in euphotic zone} - \text{DOC introduced from deeper layers} \quad \text{(Eq. 2)}$$

The ratio between ΔDOC and ΔNO_3^- gave us the NDPr:

$$NDPr = \Delta DOC / \Delta NO_3^- \quad \text{(Eq. 3)}$$

NDPr was calculated throughout the Atlantic Ocean using observations of DOC and NO_3^- from >15 international oceanographic cruises over the last decade, including those occupied by the US Repeat Hydrography program. Values of NDPr mostly varied between 0.1 and 0.4, with the exception of the North Atlantic Subtropical Gyre (NASG), where NDPr values reach >0.8 at times. After sensitivity testing, we applied a NDPr value of 0.17 to the entire basin, which yielded the smallest error between calculated and observed DOC concentrations. Applying this NDPr value to ΔNO_3^- (i.e. NCP) obtained from cruise data, we estimate ÄDOC (Eq. 4), in which 6.6 is the molar conversion from N to C units:

$$\Delta DOC = \Delta NO_3^- * 6.6 * 0.17 = NCP * 0.17 \qquad \text{(Eq. 4)}$$

To obtain the calculated DOC concentration ($DOC_{calculated}$), we added the DOC concentration of underlying source waters (DOC_{source}) to ΔDOC (Eq. 5):

$$DOC_{calculated} = DOC_{source} + \Delta DOC \qquad \text{(Eq. 5)}$$

When comparing calculated vs. observed DOC, we found significant agreement ($R^2 = 0.64$; $p < 0.001$; n=268) throughout the Atlantic, except in the western North Atlantic, where observed DOC > estimated DOC, especially in the southern sector. After this validation of our approach using nutrients and DOC observations, we applied the method to the more extensive NO_3^- distributions available in the World Atlas Ocean (WOA) climatology to develop a $DOC_{calculated}$ map for the entire Atlantic. The calculated values agree well with the observations, with a total error of 8.94%.

How much DOC is annually produced in the surface Atlantic Ocean? Total organic carbon export (considered equivalent to NCP) in the Atlantic has been estimated to be 4.15-4.3 Pg C y^{-1}. Applying the 0.17 NDPr (eq. 3) indicates that 0.70-0.75 Pg C y^{-1} accumulates in the Atlantic surface as DOC; as such, the Atlantic accounts for ~36% of the global net DOC production ~2 Pg C y^{-1}.

In permanently stratified areas like the southern sectors of the NASG, our approach is invalid since there is little nutrient input from underlying depths. Also, the static view of our approach does not take into account advection that will modify the DOC distributions, nor does it account for eventual removal of accumulated and advected DOC by microbes. To account for these influences on distributions, we applied the ΔNO_3^- measurements to a steady-state ocean circulation model including terrestrial DOC inputs and DOC remineralization. In the model, zonal advection is evident through enrichment of DOC in the Caribbean Sea. Also, inputs of terrestrial DOC are observed near the outflow of the Amazon River. However, the model only slightly improved the match between observations and modeled DOC, with a total error of 8.71% vs. the 8.94% obtained before the model application.

The correspondence between observations and modeled values was good, considering that we are comparing observations of DOC from cruises during specific seasons with estimates based on more idealized nutrient climatology. The main mismatch is found in the western NASG, where observations can reach 13 μmol kjg-1 higher than calculated values. Local production and/or allochthonous inputs of either new nutrients or DOC must be considered. Local production of DOC could result from addition of nitrogen from sources beyond vertical mixing such as diazotrophic N_2 fixation, atmospheric deposition, and river runoff. Alternatively, DOC can be concentrated by evaporation, as is sea salt.

However, none of these explain the high DOC values observed in the NASG. DOC flux estimated from dissolved organic nitrogen (DON) released by N_2 fixation is too low to explain the extra DOC. Regarding the atmospheric deposition, aerosol optical depth data suggest higher deposition in the eastern than in the western North Atlantic, and no excess of DOC is observed there. According to salinity distributions from the World Ocean Atlas, advection of DOC from the closest major rivers (Amazon and Orinoco) does not extend far enough northward to explain the NASG anomaly. Salinity normalization of DOC does not erase the feature, indicating that evaporation is not the cause. Those elevated values of carbon are found during cruises from 2003 in the same area, so it appears to be a persistent feature. The anomaly also coincides with a DON maximum and a light stable isotope ($\Delta^{15}N$) composition in the particulate organic carbon based on measurements recorded in 2004. An explanation for these anomalies has not been confirmed.

CONTROL OF DISTRIBUTION OF OXYGEN IN THE OCEAN

Oxygen – product and elixir of life

Carbon dioxide, which occurs in relatively small amounts in the atmosphere, is both a crucial substance for plants, and a climate-threatening gas. Oxygen, on the other hand, is not only a major component of the atmosphere, it is also the most abundant chemical element on Earth. The emergence of oxygen in the atmosphere is the result of a biological success model, photosynthesis, which helps plants and bacteria to convert inorganic materials such as carbon dioxide and water to biomass. Oxygen was, and continues to be generated by this process. The biomass produced is, for its part, the nutritional foundation for consumers, either bacteria, animals or humans. These consumers cannot draw their required energy from sunlight as the plants do, rather they have to obtain it by burning biomass, a process that consumes oxygen. Atmospheric oxygen on our planet is thus a product, as well as the elixir of life.

Photosynthetic production of oxygen is limited, however, to the uppermost, sunlit layer of the ocean. This only extends to a depth of around 100 metres and, because of the stable density layering of the ocean, it is largely separated from the enormous underlying volume of the deeper ocean. Moreover, most of the oxygen generated by the primary producers escapes into the atmosphere within a short time, and thus does not contribute to oxygen enrichment in the deep water column. This is because the near-surface water, which extends down to around 100 metres, is typically saturated with oxygen by the supply from the atmosphere, and thus cannot store additional oxygen from biological production. In the inner ocean, on the other hand, there is no source of oxygen. Oxygen enters the ocean in the surface water through contact with the atmosphere. From there the oxygen is then brought to greater depths through the sinking and circulation of water masses. These, in turn, are dynamic processes that are strongly affected by climatic conditions.

Three factors ultimately determine how high the concentration of dissolved oxygen is at any given point within the ocean:

1. The initial oxygen concentration that this water possessed at its last contact with the atmosphere.
2. The amount of time that has passed since the last contact with the atmosphere. This can, in fact, be decades or centuries.
3. Biological oxygen consumption that results during this time due to the respiration of all the consumers. These range from miniscule bacteria to the zooplankton, and up to the higher organisms such as fish.

The present-day distribution of oxygen in the internal deep ocean is thus determined by a complicated and not fully understood interplay of water circulation and biological productivity, which leads to oxygen consumption in the ocean's interior. Extensive measurements have shown that the highest oxygen concentrations are found at high latitudes, where the ocean is cold, especially well-mixed and ventilated. The mid-latitudes, by contrast, especially on the western coasts of the continents, are characterized by marked oxygen-deficient zones. The oxygen supply here is very weak due to the sluggish water circulation, and this is further compounded by elevated oxygen consumption due to high biological productivity. This leads to a situation where the oxygen is almost completely depleted in the depth range between 100 and 1000 metres. This situation is also observed in the northern Indian Ocean in the area of the Arabian Sea and the Bay of Bengal.

Different groups of marine organisms react to the oxygen deficiency in completely different ways, because of the wide range of tolerance levels of different marine animals to oxygen-poor conditions. For instance, crustaceans and fish generally require higher oxygen concentrations than mussels or snails. The largest oceanic oxygen minimum zones, however, because of their extremely low concentrations, should be viewed primarily as natural dead zones for the higher organisms, and by no means as caused by humans.

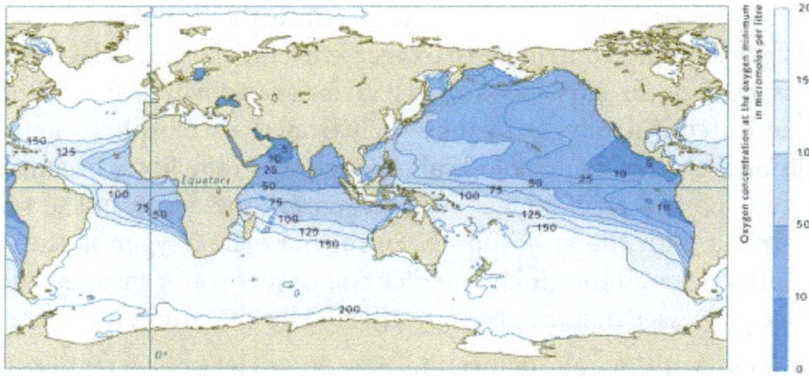

Fig. 18 Marine regions with oxygen deficiencies are completely natural. These zones are mainly located in the mid-latitudes on the west sides of the continents. There is very little mixing here of the warm surface waters with the cold deep waters, so not much oxygen penetrates to greater depths. In addition, high bioproductivity and the resulting large amounts of sinking biomass here lead to strong oxygen consumption at depth, especially between 100 and 1000 metres.

Oxygen – the renaissance of a hydrographic parameter

Oxygen distribution in the ocean depends on both biological processes, like the respiration of organisms, and on physical processes such as current flow. Changes in either of these processes should therefore lead to changes in the oxygen distribution. In fact, dissolved oxygen can be viewed as a kind of sensitive early warning system for global (climate) change in the ocean. Scientific studies show that this early warning system can detect the expected decrease in oxygen transport from the atmosphere into the ocean that is driven by global current and mixing processes, as well as possible changes in the marine biotic communities. In recent years, this knowledge has led to a renaissance of oxygen in the field of global marine research.

In oceanography, dissolved oxygen has been an important measurement parameter for over a hundred years. A method for determining dissolved oxygen was developed as early as the end of the 19th century, and it is still applied in an only slightly modified form today as a precise method. This allowed for the development of an early fundamental understanding of the oxygen distribution in the world ocean, with the help of the famous German Atlantic Expedition of the "Meteor" in the 1920s.

Research efforts in recent years have recorded decreasing oxygen concentrations for almost all the ocean basins. These trends are, in part, fairly weak and mainly limited to water masses in the upper 2000 metres of the ocean. Therefore, no fully consistent picture can yet be drawn from the individual studies. Most of the studies do, however, show a trend of decreasing oxygen concentrations. This trend agrees well with an already verified expansion and intensification of the natural oxygen minimum zones, those areas that are deadly for higher organisms. If the oxygen falls below certain (low) threshold values, the water becomes unsuitable for higher organisms. Sessile, attached organisms die. Furthermore, the oxygen deficiency leads to major changes in biogeochemical reactions and elemental cycles in the ocean – for instance, of the plant nutrients nitrate and phosphate.

Oxygen levels affect geochemical processes in the sediment but also, above all, bacterial metabolism processes, which, under altered oxygen conditions, can be changed dramatically. It is not fully possible today to predict what consequences these changes will ultimately have. In some cases it is not even possible to say with certainty whether climate change will cause continued warming, or perhaps even local cooling. But it is probable that the resulting noticeable effects will continue over a long time period of hundreds or thousands of years

Even today, however, climate change is starting to cause alterations in the oxygen content of the ocean that can have negative effects. For the first time in recent years, an extreme low-oxygen situation developed off the coast of Oregon in the United States that led to mass mortality in crabs and fish. This new death zone off Oregon originated in the open ocean and presumably can be attributed to changes in climate. The prevailing winds off the west coast of the USA apparently changed direction and intensity and, as a result, probably altered the ocean currents. Researchers believe that the change caused oxygen-poor water from greater depths to flow to surface waters above the shelf.

The death zone off Oregon is therefore different than the more than 400 near-coastal death zones known worldwide, which are mainly attributed to eutrophication, the excessive input of plant nutrients. Eutrophication normally occurs in coastal waters near densely populated regions with intensive agricultural activity.

Oxygen – challenge to marine research

The fact that model calculations examining the effects of climate change almost all predict an oxygen decline in major parts of the ocean, which agrees with the available observations of decreasing oxygen, gives the subject additional weight. Even though the final verdict is not yet in, there are already indications that the gradual loss of oxygen in the world ocean is an issue of great relevance which possibly also has socio-economic repercussions, and which ocean research must urgently address.

Intensified research can provide more robust conclusions about the magnitude of the oxygen decrease. In addition it will contribute significantly to a better understanding of the effects of global climate change on the ocean. In recent years marine research has addressed this topic with increased vigour, and has already established appropriate research programmes and projects. It is difficult, however, to completely measure the temporally and spatially highly variable oceans in their totality. In order to draw reliable conclusions, therefore, the classic instruments of marine research like ships and taking water samples will not suffice. Researchers must begin to apply new observational concepts.

"Deep drifters" are an especially promising tool: these are submersible measuring robots that drift completely autonomously in the ocean for 3 to 4 years, and typically measure the upper 2000 metres of the water column every 10 days. After surfacing, the data are transferred to a data centre by satellite. There are presently around 3200 of these measuring robots deployed for the international research programme ARGO, named after a ship from Greek mythology. Together they form a worldwide autonomous observatory that is operated by almost 30 countries.

So far this observatory is only used on a small scale for oxygen measurements. But there has been developed a new sensor technology for oxygen measurements in the recent past that can be deployed on these drifters. This new technology would give fresh impetus to the collection of data on the variability of the oceanic oxygen distribution.

CHAPTER

AIR-SEA INTERACTION 6

INTRODUCTION

Air-sea interaction describes the dynamic, thermodynamic, and biogeochemical processes that link the atmosphere and oceans. The ocean stores incoming energy from the sun and then transfers heat and moisture to the atmosphere that fuels its circulation. These atmospheric circulations result in our weather systems that include mid-latitude storms and hurricanes. Once in motion, the atmosphere drives additional ocean circulation through the exchange of momentum across the air-sea interface that drives waves and currents. Air-sea interaction also regulates the storage and release of anthropogenic and naturally occurring greenhouse gases, the generation of aerosols and cloud condensation nuclei from both sea-salt and phytoplankton-derived dimethyl sulfide. All of these processes fundamentally depend on the exchange of momentum, heat and mass across marine boundary layer.

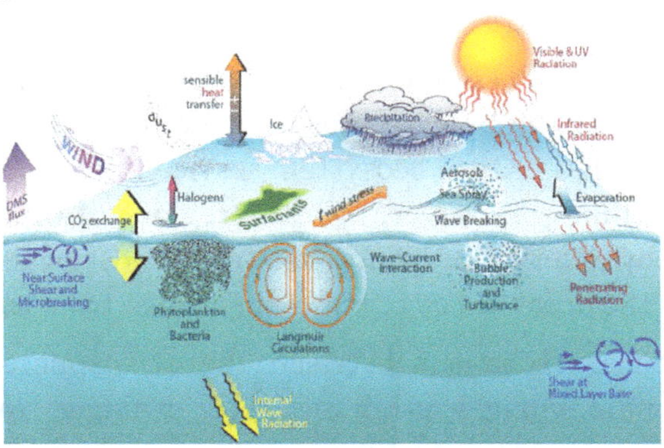

Fig. 1. Air-Sea Interaction

From the physical point of view, the interaction between these two turbulent fluids, the ocean and the atmosphere, is a complex, highly nonlinear process, fundamental to the motions of both. The winds blowing over the surface of the ocean transfer momentum and mechanical energy to the water, generating waves and currents. The ocean in turn gives off energy as heat, by the emission of electromagnetic radiation, by conduction, and, in latent form, by evaporation.

The heat flux from the ocean provides one of the main energy sources for atmospheric motions. This source of energy for the atmosphere is affected by the turbulence at the air/sea interface, and by the spatial distribution of the centres of high and low energy transfer affected by the ocean currents. This coupling takes place through processes that fundamentally occur at small scales. The strength of this coupling depends on air-sea differences in several factors and therefore has geographic and temporal scales over a broad range. At these small scales on the sea-surface interface itself, waves, winds, water temperature and salinity, bubbles, spray and variations in the amount of solar radiation that reaches the ocean surface, and other factors, affect the transfer of properties and energy.

In the long term, the convergence and divergence of oceanic heat transport provide sources and sinks of heat for the atmosphere and partly shape the mean climate of the earth. Analyzing whether these processes are changing due to anthropogenic influences and the potential impact of these changes is the subject of this chapter. Following guidance from the Ad Hoc Working Group of the Whole, much of the information presented here is based on or derives from the very thorough analysis conducted by the Intergovernmental Panel on Climate Change (IPCC) for its recent Fifth Assessment Report (AR5).

From a biogeochemical point of view, gas and chemical exchanges between the oceans and the atmosphere are important to life processes. Half of the Global Net Primary Production of the world is by phytoplankton and other marine plants, uptaking CO_2 and releasing oxygen. Phytoplankton is therefore also responsible for half of the annual production of oxygen by plants and, through the generation of organic matter, is at the basis of most marine food webs in the ocean. Oxygen production by plants is a critical ecosystem service that keeps atmospheric oxygen from otherwise declining. However, in many regions of the ocean, phytoplankton growth is limited by a deficit of iron in seawater. Most of the iron alleviating this limitation reaches the ocean through wind-borne dust from the deserts of the world.

Gas and chemical exchanges between the atmosphere and ocean are also important to climate change processes. For example, marine phytoplankton produces dimethyl sulphide (DMS), the most abundant biological sulphur compound emitted to the atmosphere. DMS is oxidized in the marine atmosphere to form various sulphur-containing compounds, including sulphuric acid, which influence the formation of clouds. Through this interaction with cloud formation, the massive production of atmospheric DMS over the ocean may have an impact on the earth's climate. The absorption of CO_2 from the atmosphere at the sea surface is responsible for the fundamental role of the ocean as a carbon sink.

WHAT CAUSES VARIATIONS IN SOLAR RADIATION ON EARTH?

While the solar radiation incident on the Earth's atmosphere is relatively constant, the radiation at the Earth's surface varies widely due to:

- atmospheric effects, including absorption and scattering;
- local variations in the atmosphere, such as water vapour, clouds, and pollution;
- latitude of the location; and
- the season of the year and the time of day.

Fig. 2. Solar radiation at the Earth's surface varies from the solar radiation incident on the Earth's atmosphere. Cloud cover, air pollution, latitude of a location, and the time of the year can all cause variations in solar radiance at the Earth's surface.

The above effects have several impacts on the solar radiation received at the Earth's surface. These changes include variations in the overall power received, the spectral content of the light and the angle from which light is incident on a surface. In addition, a key change is that the variability of the solar radiation at a particular location increases dramatically. The variability is due to both local effects such as clouds and seasonal variations, as well as other effects such as the length of the day at a particular latitude. Desert regions tend to have lower variations due to local atmospheric phenomena such as clouds. Equatorial regions have low variability between seasons.

The amount of energy reaching the surface of the Earth every hour is greater than the amount of energy used by the Earth's population over an entire year.

Solar Variability: Striking a Balance with Climate Change

The sun has powered almost everything on Earth since life began, including its climate. The sun also delivers an annual and seasonal impact, changing the character of each hemisphere as Earth's orientation shifts through the year. Since the Industrial Revolution, however, new forces have begun to exert significant influence on Earth's climate.

For the past three decades NASA scientists have investigated the unique relationship between the sun and Earth. Using space-based tools, like the Solar Radiation and Climate Experiment (SORCE), they have studied how much solar energy illuminates Earth, and explored what happens to that energy once it penetrates the atmosphere. The amount of energy that reaches Earth's outer atmosphere is called the total solar irradiance. Total solar irradiance is variable over many different timescales, ranging from seconds to centuries due to changes in solar activity.

The sun goes through roughly an 11-year cycle of activity, from stormy to quiet and back again. Solar activity often occurs near sunspots, dark regions on the sun caused by concentrated magnetic fields. The solar irradiance measurement is much higher during solar maximum, when sunspot cycle and solar activity is high, versus solar minimum, when the sun is quiet and there are usually no sunspots.

Using SORCE, scientists have learned that about 1,361 watts per square meter of solar energy reaches Earth's outermost atmosphere during the sun's quietest period. But when the sun is active, 1.3 watts per square meter (0.1 percent) more energy reaches Earth. "This TSI measurement is very important to climate models that are trying to assess Earth-based forces on climate change," said Cahalan.

Over the past century, Earth's average temperature has increased by approximately 0.6°C (1.1 °F). Solar heating accounts for about 0.15 C, or 25 percent, of this change, according to computer modeling results published by NASA Goddard Institute for Space Studies researcher David Rind in 2004. Earth's climate depends on the delicate balance between incoming solar radiation, outgoing thermal radiation and the composition of Earth's atmosphere. Even small changes in these parameters can affect climate. Around 30 percent of the solar energy that strikes Earth is reflected back into space. Clouds, atmospheric aerosols, snow, ice, sand, ocean surface and even rooftops play a role in deflecting the incoming rays. The remaining 70 percent of solar energy is absorbed by land, ocean, and atmosphere.

Greenhouse gases block about 40 percent of outgoing thermal radiation that emanates from Earth. The resulting imbalance between incoming solar radiation and outgoing thermal radiation will likely cause Earth to heat up over the next century, accelerating the melting polar ice caps, causing sea levels to rise and increasing the probability of more violent global weather patterns.

Non-Human Influences on Climate Change

Before the Industrial Age, the sun and volcanic eruptions were the major influences on Earth's climate change. Earth warmed and cooled in cycles. Major cool periods were ice ages, with the most recent ending about 11,000 years ago.

The sun is relatively calm compared to other stars. "We don't know what the sun is going to do a hundred years from now," said Doug Rabin, a solar physicist at Goddard. "It could be considerably more active and therefore have more influence on Earth's climate."

Or, it could be calmer, creating a cooler climate on Earth similar to what happened in the late 17th century. Almost no sunspots were observed on the sun's surface during the period from 1650 to 1715. This extended absence of solar activity may have been partly responsible for the Little Ice Age in Europe and may reflect cyclic or irregular changes in the sun's output over hundreds of years. During this period, winters in Europe were longer and colder by about 1°C than they are today.

Since then, there seems to have been on average a slow increase in solar activity. Unless we find a way to reduce the amount of greenhouse gases we put into the atmosphere, such as carbon dioxide from fossil fuel burning, the solar influence is not expected to dominate climate change. But the solar variations are expected to continue to modulate both warming and cooling trends at the level of 0.1 to 0.2 degrees Celsius (0.18 to 0.26 Fahrenheit) over many years.

FUTURE MEASUREMENTS OF SOLAR VARIABILITY

For three decades, a suite of NASA and European Space Agency satellites have provided scientists with critical measurements of total solar irradiance. The Total Irradiance Monitor, also known as the TIM instrument, was launched in 2003 as part of the NASA's SORCE mission, and provides irradiance measurements with state-of-the-art accuracy. TIM has been rebuilt as part of the Glory mission, scheduled to launch in 2009. Glory's TIM instrument will continue an uninterrupted 30-year record of solar irradiance measurements and will help researchers better understand the sun's direct and indirect effects on climate. Glory will also collect data on aerosols, one of the least understood pieces of the climate puzzle.

HOW DOES THE CORIOLIS EFFECT INFLUENCE MOVING OBJECTS?

The Coriolis effect is an "apparent" effect, an illusion produced by a rotating frame of reference. This type of effect is also known as a fictitious force or an inertial force. The Coriolis effect occurs when an object moving along a straight path is viewed from a non-fixed frame of reference. Typically, this moving frame of reference is the Earth, which rotates at a fixed speed. When you view an object in the air that is following a straight path, the object will appear to lose its course because of the rotation of the Earth. The object is not actually moving off its course. It only appears to be doing so because the Earth is turning beneath it.

What is Coriolis Force?

Coriolis force is the invisible force that appears to deflect the objects. The Coriolis force is determined by the object's rate of rotation and mass of the object.

Coriolis Effect on Earth

Coriolis force is perpendicular to the object's axis. As the Earth spins on its axis from west to east, the Coriolis force acts north-south. At the Equator, the Coriolis force is zero.

Causes of the Coriolis Effect

The main cause of the Coriolis effect is the Earth's rotation. As the Earth spins in a counter-clockwise direction on its axis, anything flying or flowing over a long distance above its surface is deflected. This occurs because as something moves freely above the Earth's surface, the Earth moves east under the object at a faster speed.

Impacts of the Coriolis Effect

Some of the most important impacts of the Coriolis effect in terms of geography are the deflection of winds and currents in the ocean. There is also a significant effect on man-made items such as planes and missiles.

In terms of affecting the wind, as air rises off of the Earth's surface, its speed over the surface increases because there's less drag as the air no longer has to move across the Earth's many types of landforms. Because the Coriolis effect increases with an object's increasing speed, it significantly deflects air flows.

In the Northern Hemisphere these winds spiral to the right and in the Southern Hemisphere they spiral to the left. This usually creates the westerly winds moving from the subtropical areas to the poles.

Because currents are driven by the movement of wind across the waters of the ocean, the Coriolis effect also affects the movement of the ocean's currents. Many of the ocean's largest currents circulate around warm, high-pressure areas called gyres. The Coriolis effect creates the spiraling pattern in these gyres.

Finally, the Coriolis effect is important to man-made objects as well, especially when they travel long distances over the Earth. Take, for example, a flight leaving from San Francisco, California, that is heading to New York City. If the Earth did not rotate, there would be no Coriolis effect and thus the pilot could fly in a straight path to the east. However, due to the Coriolis effect, the pilot has to constantly correct for the Earth's movement beneath the plane. Without this correction, the plane would land somewhere in the southern portion of the United States.

Demonstration of Coriolis Effect

Let us pretend to be standing at the North Pole and tossing a ball to our friend standing at the equator. While the ball travels through the air, the Earth below it is rotating. Hence, when the ball reaches the equator, it lands in a location somewhere to the west of where you were aiming.

Near the earth's surface, the Coriolis effect creates wind (and water) patterns that move to the east toward the equator and to the west toward the poles. These prevailing wind patterns are responsible for moving clouds around the globe and, thus, creating weather patterns in different regions.

Characteristics of Coriolis Effect

The Coriolis effect characteristics can be summarised as follows:

- Coriolis force is a fictitious force resulting from the rotational movement of the earth.
- Coriolis effect is effective on objects that are in motion such as wind, aircraft, ballistic and flying birds.
- Coriolis effect only affects the wind direction and not the wind speed as it deflects the wind direction from the expected path.
- The magnitude of Coriolis force is determined by wind speed. The higher the wind speed, the greater the deflection.
- Coriolis effect is maximum at the poles and zero at the equator.
- Coriolis force always acts in a direction that is perpendicular to the moving object's axis.

HOW ARE WEATHER PATTERNS AFFECTED BY THE CORIOLIS EFFECT?

The development of weather patterns, such as cyclones and trade winds, are examples of the impact of the Coriolis Effect.

In the Northern Hemisphere, fluids from high-pressure systems pass low-pressure systems to their right. As air masses are pulled into cyclones from all directions, they are deflected, and the storm system, a hurricane, seems to rotate counter-clockwise.

In the Southern Hemisphere, currents are deflected to the left. As a result, storm systems seem to rotate clockwise.

Coriolis Effect and Ocean Current

Ocean currents are the continuous, directional and predictable movement of seawater. Ocean currents are driven by the movement of wind across the ocean's waters, and the Coriolis effect dramatically affects the direction of the ocean's currents. Many of the ocean's most enormous currents circulate in warm, high-pressure areas called gyres, and the Coriolis effect creates the spiralling pattern in these gyres.

Impact of Coriolis Effect on Airplanes and Human Activity

Fast-moving objects impacted by weather, such as aeroplanes and rockets, are influenced by the Coriolis Effect. The Coriolis Effect largely determines the direction of the prevailing winds. Hence a pilot must take this into account while charting routes for long-distance travel.

Military snipers consider the Coriolis effect. Although bullets' trajectory is minimal to be significantly affected by the Earth's rotation, sniper targeting is so precise that a deflection of several centimetres could injure innocent people or damage civilian infrastructure.

GLOBAL ATMOSPHERIC CIRCULATION

The global circulation

Over the major parts of the Earth's surface there are large-scale wind circulations present. The global circulation can be described as the world-wide system of winds by which the necessary transport of heat from tropical to polar latitudes is accomplished.

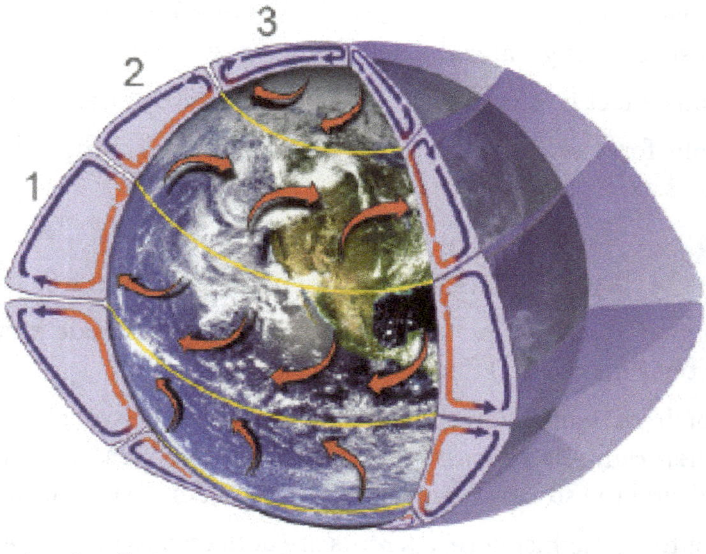

Fig. 3. Three main circulations exist between the equator and poles due to earth's rotation: 1. Hadley cell, 2. Ferrel cell and 3. Polar cell

In each hemisphere there are three cells (Hadley cell, Ferrel cell and Polar cell) in which air circulates through the entire depth of the troposphere. The troposphere is the name given to the vertical extent of the atmosphere from the surface, right up to between 10 and 15 km high. It is the part of the atmosphere where most of the weather takes place.

Hadley cell

The largest cells extend from the equator to between 30 and 40 degrees north and south, and are named Hadley cells, after English meteorologist George Hadley. Within the Hadley cells, the trade winds blow towards the equator, then ascend near the equator as a broken line of thunderstorms, which forms the Inter-Tropical-Convergence Zone (ITCZ). From the tops of these storms, the air flows towards higher latitudes, where it sinks to produce high-pressure regions over the subtropical oceans and the world's hot deserts, such as the Sahara desert in North Africa.

Ferrel cell

In the middle cells, which are known as the Ferrel cells, air converges at low altitudes to ascend along the boundaries between cool polar air and the warm subtropical air that generally occurs between 60 and 70 degrees north and south. This often occurs around the latitude of the UK which gives us our unsettled weather. The circulation within the Ferrel cell is complicated by a return flow of air at high altitudes towards the tropics, where it joins sinking air from the Hadley cell.

The Ferrel cell moves in the opposite direction to the two other cells (Hadley cell and Polar cell) and acts rather like a gear. In this cell the surface wind would flow from a southerly direction in the northern hemisphere. However, the spin of the Earth induces an apparent motion to the right in the northern hemisphere and left in the southern hemisphere. This deflection is caused by the Coriolis effect and leads to the prevailing westerly and south-westerly winds often experienced over the UK.

Polar cell

The smallest and weakest cells are the Polar cells, which extend from between 60 and 70 degrees north and south, to the poles. Air in these cells sinks over the highest latitudes and flows out towards the lower latitudes at the surface.

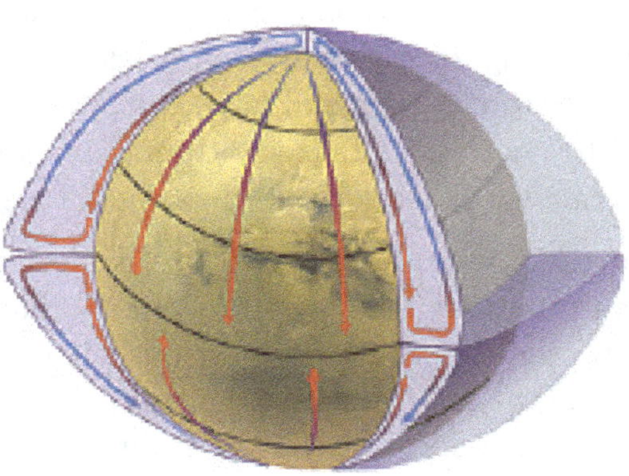

Fig. 4. Air flow for no rotation and no water on a planet.

Global Circulations explain how air and storm systems travel over the Earth's surface. The global circulation would be simple (and the weather boring) if the Earth did not rotate, the rotation was not tilted relative to the sun, and had no water.

In a situation such as this, the sun heats the entire surface, but where the sun is more directly overhead it heats the ground and atmosphere more. The result would be the equator becomes very hot with the hot air rising into the upper atmosphere.

That air would then move toward the poles where it would become very cold and sink, then return to the equator (Fig. 3). One large area of high pressure would be at each of the poles with a large belt of low pressure around the equator.

However, since the earth rotates, the axis is tilted, and there is more land mass in the northern hemisphere than in the southern hemisphere, the actual global pattern is much more complicated.

HOW DOES THE OCEAN INFLUENCE GLOBAL WEATHER PHENOMENA AND CLIMATE PATTERNS?

Covering about 70 percent of the Earth's surface, the world's oceans have a two-way relationship with weather and climate. The oceans influence the weather on local to global scales, while changes in climate can fundamentally alter many properties of the oceans. This section examines how some of these important characteristics of the oceans have changed over time.

How the oceans influence climate

The oceans influence climate over long and short time-scales. On the longest time-scale of geologic time, the shape and location of the continents helps to determine the oceans' circulation patterns. Since continental plates drift at about 5 cm per year and mountain ranges rise by about 1 mm, it usually takes millions of years for new land formations to change the oceans. Patterns of ocean circulation and up-welling can also change much more rapidly, resulting in climate variations and fluctuations on a human time-scale. Records of global and, in particular, regional climate show periods lasting from years to centuries during which the climate was systematically different from earlier and later periods. Scientists believe that this behaviour is related to changes in the way the oceans store and transport heat, although the precise causes of these changes are not always clear.

The oceans and the atmosphere are tightly linked and together form the most dynamic component of the climate system. Changes in external factors such the sun's energy, the distribution of various plant species, or the emission of greenhouse gases into the atmosphere can alter the temperature and circulation patterns of the atmosphere-ocean system. Because the atmosphere and oceans are turbulent, they can also generate their own internal fluctuations. Short-term fluctuations in wind or temperature (that is, weather) can directly influence the currents and temperature of the underlying ocean, while oceanic fluctuations can magnify, diminish, or modify atmospheric fluctuations.

The oceans play a critical role in storing heat and carbon. When the earth's surface cools or is heated by the sun, the temperature change is greater - and faster - over the land than over the oceans. Because it is a fluid, the ocean diffuses the effects of a temperature change for great distances via vertical mixing and convective movements. The solid land cannot, so the sun's heat penetrates only the thin, upper crust. One consequence of the ocean's ability to absorb more heat is that when an area of ocean becomes warmer or cooler than usual, it takes much longer for that area to revert to "normal" than it would for a land area. This also explains why "maritime" climates tend to be less extreme than "continental" ones, with smaller day-night and winter-summer differences.

The ocean's waters are constantly being moved about by powerful currents. Surface currents are largely wind-driven, although the rotation of the earth, the presence of continents, and the oceans' internal dynamics also have a strong influence. Deep-ocean flow (and, to a lesser extent, surface flow) is driven by density differences produced by heating and cooling and by precipitation and evaporation (cool saline water is denser than warm fresh-water). The behaviour of the atmosphere strongly affects these density differences. For example, clouds can cool the sea by blocking the warming rays of the sun or reduce surface salinity by bringing rain. The wind can influence evaporation rates by blowing more strongly or more weakly.

These currents influence the climate by transporting heat. Horizontal currents, particularly those moving north or south, can carry warmed or cooled water as far as several thousand kilometres. The displaced water can then warm or cool the air and, indirectly, the land over which this air blows. For example, water from the tropical and subtropical Atlantic (including some from the Gulf of Mexico) moves north through the Atlantic in a current popularly (if misleadingly) called the "Gulf Stream". There it bathes the shores of Western Europe, producing a climate that is surprisingly mild for that latitude. In addition to currents, up-wellings of cold water in places where the wind blows surface water away can also affect climate. Thus San Francisco, influenced by coastal up-welling, is hardly warmer than Dublin, which is influenced by the Gulf Stream, despite being over 1,600 km further south.

Currents involved in "deep-water formation" are particularly important for climate. In winter, surface cooling causes water to become more dense. (While fresh-water that is cooled starts to expand at temperatures below 4°C, salt-water continues to compress all the way down to its freezing point of -2°C.) In areas where evaporation exceeds precipitation, the resulting rise in salinity also increases density.

When the surface water becomes denser than the underlying water, "convective overturning" occurs and the dense surface water mixes downwards. In certain places this downward mixing can occasionally extend all the way to the bottom, even in deep oceans. The dense, deep water thus formed spreads out over the whole ocean. As a result, when downward mixing takes place at high latitudes it creates a circulation pattern in which warm water from tropical and subtropical regions moves poleward, surrenders heat to the atmosphere, cools and sinks, and flows back towards the equator. The net result is a transport of heat poleward.

An apparently small change in just one aspect of the ocean's behaviour can produce major climate variations over large areas of the earth. The areas of cold-water formation are one known example of this possibly wide-spread phenomenon. Although more research is needed, there is some agreement among oceanographers that, for the entire area north of 30 N latitude, the ocean's poleward transport of heat is the equivalent of about 15 watts per square metre of the earth's surface (W/m2). This can be compared with some 200 W/m2 from direct sunshine, and about 6 W/m2 for what climate change models predict will happen if the atmospheric concentration of carbon dioxide doubles. Recent observations, ocean core records, and some modelling results indicate that North Atlantic deep-water formation and its associated ocean heat flow fluctuate substantially over time-scales ranging from years to millennia. The system is vulnerable because even a relatively small decrease in surface salinity prevents water - no matter how cold it is - from sinking. This could occur if there is a flood of fresh-water run-off from the Arctic due to global warming.

Environmental, economic and social impacts of changes in ocean temperature and of major ocean temperature events

Coastal waters are valuable both ecologically and economically because they support a high level of biodiversity. They act as nursery areas for many commercially important fish species, and are the marine areas most accessible to the public. Because inshore habitats are shallow, water temperatures in coastal areas are closely linked to the regional climate and its seasonal and long-term fluctuations.

Coastal waters also host some of the most vulnerable marine habitats, because they are intensively exploited by (including, but not limited to) the fishing industry and recreational craft, and because of their proximity to outlets of pollution, such as rivers and sewage outfalls.

Coastal development and the threat of rising sea level may also impinge upon these valuable habitats (Halpern et al., 2008). Ecological degradation can lower the socio-economic value of coastal regions, with negative impacts on commercial fisheries, aquaculture facilities, damage to coastal infrastructure, problems with power-station cooling, and exert a dampening effect on coastal tourism from degraded ecological services.

It has been recently shown that when compared with estimates for the global ocean, decadal rates of SST change are higher at the coast. During the last three decades, approximately 70 per cent of the world's coastline has experienced significant increases in SST.

This has been accompanied by an increase in the number of yearly extremely hot days along 38 per cent of the world's coastline, and warming has been occurring significantly earlier in the year along approximately 36 per cent of the world's temperate coastal areas (defined as those between latitudes 30° and 60° in both hemispheres) at an average rate of 6.1 ± 3.2 days per decade.

The warming of coastal waters can have many serious consequences for the ecological system. This can include changes in the distribution of important commercial fish and shellfish species, particularly the movement of species to higher latitudes due to thermal stress. Warming of coastal waters also can lead to more favourable conditions for many organisms, among them marine invasive species that can devastate commercial fisheries and destroy marine ecosystem dynamics.

Water quality might also be impacted by higher temperatures that can increase the severity of local outbreaks by pathogenic bacteria or the occurrence of Harmful Algal Blooms (HABs). These in turn would cause harm to seafood, consumers and marine organisms. Increased coral reef bleaching and mortality from warming seas (combined with ocean acidification, see next sections) will lead to the loss of important marine habitats and associated biodiversity.

Changes in ocean temperatures have global impacts. As ocean temperatures warm, species that prefer specific temperature ranges may relocate – as has been observed, for instance, in copepod assemblages in the North Atlantic. Some organisms, like corals, are sedentary and cannot relocate with changing temperatures. If the water becomes too warm, they may experience a bleaching event. Higher sea level and warmer ocean temperatures can alter ocean circulation and current flow and increase the frequency and intensity of storms, leading to changes in the habitat of many species worldwide.

Changes in ocean temperatures affect not only marine ecosystems, but also the climate over land, with devastating economic and social implications. Many natural oceanic oscillations are known to have an impact on (terrestrial) climate, but these oscillations and the response of the climate to them are also changing during recent decades. For instance, an El Niño phase of ENSO displaces great amounts of warm water from the western to the eastern Pacific, leading to more evaporation over the latter. As a consequence, western and southern South America and parts of North America experience wetter conditions. At the same time, Australia, Brazil, India, Indonesia, the Philippines, parts of Africa and the United States of America suffer droughts. La Niña events usually cause the opposite patterns. However, in the last several decades, ENSO events have changed their spatial and temporal characteristics.

During recent decades, the warm waters of El Niño events have been displaced to the central Pacific instead of to the eastern Pacific. It is not clear yet whether these changes are linked to anthropogenic climate change or natural variability. In any case, the effects on climate of an ENSO event centred in the central Pacific (a central Pacific ENSO) are in sharp contrast to that associated with one centred in the eastern Pacific.

For instance, northeastern and southeastern Australia experience a reduction in rainfall during the eastern Pacific El Niños and there is a decrease in rainfall over northwestern and northern Australia during central Pacific events. The Indian monsoon fails during eastern Pacific El Niños, but is enhanced during central Pacific El Niños. Over the semi-arid region of northeast Brazil, eastern Pacific El Niños/La Niñas cause dry/wet conditions; central Pacific El Niños have the opposite effect, with the worst drought in the last 50 years associated with the strong 2011/12 La Niña and not with El Niños as in the past. This drought caused the displacement of 10 million people and economic losses on the order of 3 billion United States dollars in relation to agriculture and cattle raising alone. In contrast to drought in Brazil, the 2011/12 La Niña caused floods across southeastern Australia.

In other ocean basins, changes in oceanic oscillations and temperatures have also had an impact on climate. For instance, in the Indian Ocean, a positive phase of the Indian Dipole Mode (warm/cold temperatures in the western/eastern equatorial Indian Ocean) leads to flooding in east Africa and droughts in Indonesia, Australia, and India. The counterpart of ENSO in the Atlantic (Atlantic Niño) has weakened during the last six decades, leading to an increase in SST in the eastern equatorial Atlantic. As a consequence, rainfall has been enhanced over the equatorial Amazon and West Africa. On the other hand, an unusual warming of the tropical North Atlantic in 2005 was responsible for one of the worst droughts in the Amazon River basin and a record Atlantic hurricane season. Hurricanes Rita and Katrina caused the loss of almost 2000 lives and an estimated economic toll of 150 billion —135 billion US dollars from Katrina and 15 billion US dollars from Rita. Anomalous warm conditions also occurred in the tropical North Atlantic in 2010 leading to two once-in-a-century droughts in less than five years in the Amazon River basin.

Ocean warming will stress species both through thermic changes in their environmental envelope and through increased interspecies competition. These shifts become all the more important in shelf seas once they reach terrestrial boundaries, i.e., the shifting species runs out of shelf. For example, changes in the coastal currents in south-eastern Australia cause changes to primary production through to fisheries productivity. This then feeds through to local and regional socio-economic impacts

The IPCC AR5 concluded that "it is unlikely that annual numbers of tropical storms, hurricanes and major hurricanes counts have increased over the past 100 years in the North Atlantic basin. Evidence, however, is for a virtually certain increase in the frequency and intensity of the strongest tropical cyclones since the 1970s in that region".

Moreover, the IPCC AR5 states that "it is difficult to draw firm conclusions with respect to the confidence levels associated with observed trends prior to the satellite era and in ocean basins outside of the North Atlantic". Although a strong scientific consensus on the matter does not exist, there is some evidence supporting the hypothesis that global warming might lead to fewer but more intense tropical cyclones globally. Evidence exists that the observed expansion of the tropics since approximately 1979 is accompanied by a pronounced poleward migration of the latitude at which the maximum intensities of storms occur at a rate of 1° of latitude per decade (Kossin et al., 2014; Hartmann et al., 2013; Seidel et al., 2008). If this trend is confirmed, it would increase the frequency of events in coastal areas that are not exposed regularly to the dangers caused by cyclones. Hurricane Sandy in 2012 may be an example of this.

WATER FLUX AND SALINITY

Regional patterns of salinity, and changes in salinity and freshwater content

The ocean plays a pivotal role in the global water cycle: about 85 per cent of the evaporation and 77 per cent of the precipitation occur over the ocean. The horizontal salinity distribution of the upper ocean largely reflects this exchange of freshwater: high surface salinity is generally found in regions where evaporation exceeds precipitation, and low salinity is found in regions of excess precipitation and runoff. Ocean circulation also affects the regional distribution of surface salinity.

The Earth's water cycle involves evaporation and precipitation of moisture at the Earth's surface. Changes in the atmosphere's water vapour content provide strong evidence that the water cycle is already responding to a warming climate. Further evidence comes from changes in the distribution of ocean salinity. Diagnosis and understanding of ocean salinity trends are also important, because salinity changes, like temperature changes, affect circulation and stratification, and therefore the ocean's capacity to store heat and carbon as well as to change biological productivity.

Seawater contains both salt and fresh water, and its salinity is a function of the weight of dissolved salts it contains. Because the total amount of salt does not change over human time scales, seawater's salinity can only be altered—over days or centuries—by the addition or removal of fresh water.

The water cycle is expected to intensify in a warmer climate. Observations since the 1970s show increases in surface and lower atmospheric water vapour (Fig. 5a), at a rate consistent with observed warming. Moreover, evaporation and precipitation are projected to intensify in a warmer climate. Recorded changes in ocean salinity in the last 50 years support that projection

The atmosphere connects the ocean's regions of net fresh water loss to those of fresh water gain by moving evaporated water vapour from one place to another. The distribution of salinity at the ocean surface largely reflects the spatial pattern of evaporation minus precipitation (Fig. 5b), runoff from land, and sea ice processes. There is some shifting of the patterns relative to each other, because of the ocean's currents. Ocean salinity acts as a sensitive and effective rain gauge over the ocean. It naturally reflects and smoothes out the difference between water gained by the ocean from precipitation, and water lost by the ocean through evaporation, both of which are very patchy and episodic. Data from the past 50 years show widespread salinity changes in the upper ocean, which are indicative of systematic changes in precipitation and runoff minus evaporation.

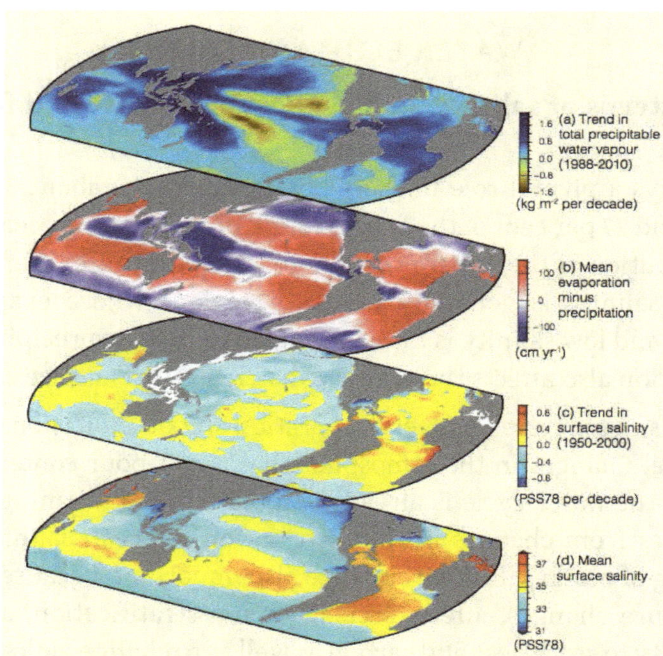

Fig. 5. The boundaries and names shown and the designations used on this map do not imply official endorsement

Subtropical waters are highly saline, because evaporation exceeds rainfall, whereas seawater at high latitudes and in the tropics—where more rain falls than evaporates—is less so. The Atlantic, the saltiest ocean basin, loses more freshwater through evaporation than it gains from precipitation, while the Pacific is nearly neutral, i.e., precipitation gain nearly balances evaporation loss, and the Southern Ocean is dominated by precipitation. (Fig. 5b). Changes in surface salinity and in the upper ocean have reinforced the mean salinity pattern (Fig. 5c). The evaporation-dominated subtropical regions have become saltier, while the precipitation-dominated subpolar and tropical regions have become fresher. When changes over the top 500 m are considered, the evaporation-dominated Atlantic has become saltier, while the nearly neutral Pacific and precipitation-dominated Southern Ocean have become fresher (Fig. 5d)

GLOBAL WARMING AND OCEAN WARMING

The ocean is on average more than two miles deep. Measuring water temperature at the surface, where it's warmed by the sun and air, gives only part of the picture.

When climate scientists calculated how much the levels of carbon dioxide in the atmosphere should heat up the air, their results didn't match observed warming: air temperatures should have been higher than they were. Where was the missing heat? It turns out that the ocean has absorbed it, like a bucket of water absorbs heat when left out in the Sun. To find it, researchers had to look not just at the surface, but half a mile down into the sea.

Water temperature at the surface of the ocean is an important driver of weather, but the temperature farther down has long-term effects on the planet. Water holds onto heat energy—that bucket of water warmed by the Sun stays warm long after the evening air has cooled down.

The ocean, by absorbing excess heat from the air, has saved us from worse warming on land, but the heat is now affecting oceanic systems.

Effects of a Warming Ocean

The rising heat content of the ocean is responsible for most of the sea level rise that coastal communities are experiencing. This is because water expands as it heats up, so a set amount of water fills more of the world's ocean basins.

Warm water also affects ice in the polar regions, melting and thinning ice shelves and sea ice (floating ice). In the last several decades, the extent of this floating ice in the Arctic Ocean has steadily dropped—a change visible in satellite views of the region.

What was once an extreme is now a norm, and climate scientists predict that by mid-century, the Arctic will be completely free of ice during the summer.

This visualization shows the extent of Arctic sea ice between 1984 and 2016. White shows older, thicker ice, while the blue-gray color is first-year ice.

Because of the heat-holding properties of water, the ocean will hold onto the increased heat for a long time, moderating the immediate effects of climate change on land but extending the impacts farther into the future.

Tracking Ocean Temperatures

To determine ocean temperatures, researchers look at data from two main sources. Orbiting satellites record the height of the ocean's surface using radar. Knowing the sea level, they can determine the water temperature, because water expands a certain amount with every degree of temperature increase. In addition, a fleet of more than 3500 floating sensors deployed across the world's oceans measure temperatures directly. Having data from more than one source lets researchers cross-check their results, giving them a clearer, more robust understanding of ocean temperature.

Ocean Acidification

Growing levels of CO_2 in the atmosphere are affecting ocean chemistry. This graph shows changes in ocean pH (acidity) over a -28year period from a monitoring station in Hawaii

How CO_2 Affects the Sea

At the surface of the ocean, there is a constant interchange between the air and the water. Water molecules evaporate and enter the air, and molecules from the air dissolve into the water—molecules including oxygen and carbon dioxide. Once in the water, carbon dioxide reacts with seawater to produce carbonic acid, CH_2CO_3. Much of the CO_2 pumped into our atmosphere by burning fossil fuels has been absorbed into the ocean, causing its chemistry to change: the water is moving from the alkaline end of the pH scale toward the acidic end. Scientists estimate that since 1900, surface waters have seen a 0.1 drop in pH, which translates into a %30 increase in acidity.

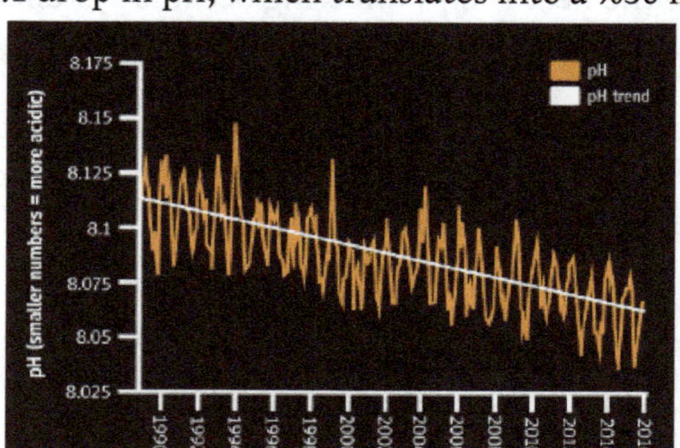

Fig. 6. Ocean pH levels at this station rise and fall seasonally (orange line), but overall they show a trend of falling pH, or growing acidity (white line).

Marine Life and the New Chemistry

The ocean's changing chemistry has important effects on many marine species. Marine animals with calcium-based shells or structures—including shellfish, corals, and some plankton—need carbonate, a molecule made of carbon and hydrogen, to make their shells. When CO_2 mixes with seawater to form carbonic acid, it uses up carbonate, reducing the availability of the molecule in the water. With less of this critical building block, marine animals can't build or maintain strong shells.

CHAPTER

OCEAN CIRCULATION 7

INTRODUCTION

Ocean currents are continuous, directed movements of ocean water generated by the forces acting upon this mean flow, such as breaking waves, wind, Coriolis force, temperature and salinity differences and tides caused by the gravitational pull of the Moon and the Sun. Depth contours, shoreline configurations and interaction with other currents influence a current's direction and strength.

Ocean currents can flow for great distances, and together they create the great flow of the global conveyor belt which plays a dominant part in determining the climate of many of the Earth's regions. Perhaps the most striking example is the Gulf Stream, which makes northwest Europe much more temperate than any other region at the same latitude. Another example is the Hawaiian Islands, where the climate is cooler (sub-tropical) than the tropical latitudes in which they are located, due to the effect of the California Current.

Ocean circulation can be conceptually divided into two main components: a fast and energetic wind-driven surface circulation, and a slow and large density-driven circulation which dominates the deep sea.

Wind-driven circulation is by far the most dynamic. Blowing wind produces currents at the surface of the ocean which are oriented at 90° to its direction (on its right in the Northern Hemisphere and on its left in the Southern Hemisphere) due to the Earth rotation. As a consequence, it creates zones of convergence or divergence of ocean currents at the point where they meet. Divergence of currents will create an upwelling phase (interior waters reach the surface) and convergence a downwelling phase (surface waters sink in the interior ocean), linking surface and interior waters.

The slow and deep circulation is largely driven by water density, and thus its temperature and salinity. It acts on the ocean as a whole and has a major influence on the abyssal properties where wind-driven circulation has no effect. However, this circulation is slow and generates weak currents, it is therefore more difficult to observe: a single drop of water travels 1,000 years to close the global overturning circulation.

The large-scale circulation is relatively stable on long timescales. At some very specific locations – mainly in the Northern Atlantic and around Antarctica – surface waters become denser and sink to the depths. Densification occurs due to both cooling surface waters and increasing salinity, the latter as a result of the removal of freshwater and the formation of ice. Surface waters are then pulled up to replace the sinking ones. How waters upwell from depths to the surface is still unclear. As stated above, zones of divergence of waters are of critical importance for these phenomena but near-seafloor turbulence also plays a major role. These mechanisms are still poorly understood and their spatial variability remains largely unknown.

Oceanic circulation is very sensitive to the global freshwater flux. This flux can be described as the difference between [Evaporation + Sea Ice Formation], which enhances salinity, and [Precipitation + Runoff + Ice melt], which decreases salinity. Global warming will undoubtedly lead to more ice melting in the poles and thus larger additions of freshwaters in the ocean at high latitudes. This input of freshwater, by decreasing surface water density near the poles, could limit downwelling, prevent deep waters formation, slowing down global circulation.

Such a process could have tremendous consequences for our societies. It would mean less carbon and heat uptake by the ocean and thus higher rates of both carbon and heat in the atmosphere. It could potentially accelerate global warming and enhance its negative effects.

More generally, it is important to note that interactions between oceanic circulation and climate are still poorly understood: more observations, an increased understanding and reliable numerical models of oceanic circulation are needed at different space and time scales. Such progress could dramatically improve IPCC global climate projections.

- Ocean currents are the continuous, predictable, directional movement of seawater. It is a massive movement of ocean water that is caused and influenced by various forces. They are like river flows in oceans.
- Ocean water moves in two directions: horizontally and vertically.
  - Horizontal movements are referred to as currents, while vertical changes are called upwellings or downwellings.
- Ocean currents impact humankind and the biosphere due to their influence on climate.

WHICH ARE THE FACTORS THAT INFLUENCES OCEAN CURRENT?

- Ocean currents are influenced by two types of forces namely:
- **Primary forces:**
  - **Heating by solar energy:** Heating by solar energy causes the water to expand. That is why, near the equator the ocean water is about 8 cm higher in level than in the middle latitudes. This causes a very slight gradient and water tends to flow down the slope.
  - **Wind:** Wind blowing on the surface of the ocean pushes the water to move. Friction between the wind and the water surface affects the movement of the water body in its course.
  - **Gravity:** Gravity tends to pull the water down the pile and create gradient variation.
  - **Coriolis force.** The Coriolis force intervenes and causes the water to move to the right in the northern hemisphere and to the left in the southern hemisphere.
  - These large accumulations of water and the flow around them are called Gyres.
  - These produce large circular currents in all the ocean basins.
- **Secondary forces:**
  - **Differences in water density:** It affects vertical mobility of ocean currents.
  - Water with high salinity is denser than water with low salinity and in the same way cold water is denser than warm water.
  - Denser water tends to sink, while relatively lighter water tends to rise.
  - **Temperature of water:** Cold-water ocean currents occur when the cold water at the poles sinks and slowly moves towards the equator.
  - Warm-water currents travel out from the equator along the surface, flowing towards the poles to replace the sinking cold water.

WHAT ARE THE TYPES OF OCEAN CURRENTS?

- The ocean currents may be classified based on their depth:
  - **Surface currents:** Large-scale surface ocean currents are driven by global wind systems that are fueled by energy from the sun.
  - These currents transfer heat from the tropics to the polar regions, influencing local and global climate.

- It constitute about 10% of all the water in the ocean, these waters are the upper 400 m of the ocean.
- **Deep water currents:** Differences in water density, **resulting from the variability of water temperature (thermo)** and **salinity (haline),** also cause ocean currents. This process is known as thermohaline circulation.
- It makes up the other 90% of the ocean water.
- These waters move around the ocean basins due to variations in the density and gravity.
- Deep waters sink into the deep ocean basins at high latitudes, where the temperatures are cold enough to cause the density to increase.
- This starts' the **global conveyor belt,** a connected system of deep and surface currents that circulate around the globe on a 1000 year time span.
- This global set of ocean currents is a critical part of Earth's climate system as well as the ocean nutrient and carbon dioxide cycles.

◉ Ocean currents can also be classified based on temperature:

- **Cold currents**: It brings cold water into warm water areas. These currents are usually found on the west coast of the continents in the low and middle latitudes (true in both hemispheres) and on the east coast in the higher latitudes in the Northern Hemisphere.
- **Warm currents**: It brings warm water into cold water areas and is usually observed on the east coast of continents in the low and middle latitudes (true in both hemispheres).
- In the northern hemisphere they are found on the west coasts of continents in high latitudes.

What are the Characteristics of Ocean Currents?

◉ Major ocean currents are greatly influenced by the stresses exerted by the prevailing winds and coriolis force. The oceanic circulation pattern roughly corresponds to the earth's atmospheric circulation pattern.

◉ The air circulation over the oceans in the middle latitudes is mainly anticyclonic (more pronounced in the southern hemisphere than in the northern hemisphere). The oceanic circulation pattern also corresponds with the same.

◉ At higher latitudes, where the wind flow is mostly cyclonic, the oceanic circulation follows this pattern.

- In regions of pronounced monsoonal flow, the monsoon winds influence the current movements.
- Due to the coriolis force, the warm currents from low latitudes tend to move to the right in the northern hemisphere and to their left in the southern hemisphere.
- The oceanic circulation transports heat from one latitude belt to another in a manner similar to the heat transported by the general circulation of the atmosphere.
- The cold waters of the Arctic and Antarctic circles move towards warmer water in tropical and equatorial regions, while the warm waters of the lower latitudes move polewards.

What are the Various Ocean Currents?

- **Equatorial Currents System:** Every ocean, except the Arctic Ocean, has a North Equatorial Current, a South Equatorial Current and an Equatorial Counter Current.
  - The North and South equatorial currents flow from east to west.
- **Equatorial Counter Current:** It is located between the North and South equatorial currents and flows in opposition to them, that is, from west to east.
- **Antarctic Circumpolar Current (ACC):** The ACC is an ocean current that flows clockwise from west to east around Antarctica. An alternative name for the ACC is the West Wind Drift.
- **Humboldt or Peruvian Current:** This low-salinity current has a large marine ecosystem and serves as one of the major nutrient systems of the world.
  - Flows from the southernmost tip of Chile to northern Peru, along the west coast of South America.
- **Kurile or Oyashio Current:** This sub-arctic ocean current circulates in a counterclockwise direction.
  - It originates in the Arctic Ocean flows south via the Bering Sea in the western North Pacific Ocean.
  - It is a nutrient-rich current.
  - It collides with Kurioshio off the Japanese eastern shore to form the North Pacific Drift.

- **California Current:** It is the extension of the Aleutian Current along the west coast of North America in a southward flowing direction.
  - It is a part of North Pacific Gyre.
  - Region of strong Upwelling.
- **Labrador Current:** It flows from the Arctic Ocean towards the south and meets the warm northward moving Gulf Stream.
  - The combination of cold Labrador Current and warm Gulf Stream is known for creating one of the richest fishing grounds of the world.
- **Canary Current:** Low salinity current extending between Fram Strait and Cape Farewell.
  - It connects the Arctic directly to the North Atlantic.
  - Major freshwater sink for the Arctic.
  - It is a major contributor to sea-ice export out of the Arctic.
- **Benguela Current:** Branch of West Wind Drift of the Southern Hemisphere.
  - Eastern portion of South Atlantic Ocean Gyre.
  - Low salinity, presence of upwelling- excellent fishing zone.
- **Falkland Current:** It is a branch of Antarctic Circumpolar Current.
  - It is also known as Malvinas Current.
  - It is named after the Falkland Islands.
  - This cold current mixes with warm Brazil current and forms the Brazil-Malvinas Confluence Zone which is responsible for the region's temperate climate.
- **Northeast Monsoon Current:** Indian North Equatorial Current flows southwest and west, crossing the Equator.
- **Somali Current:** Analogous to the Gulf Stream in the Atlantic Ocean.
  - The Current is heavily influenced by monsoon.
  - Region of major upwelling system.
- **Western Australian Current:** It is also known as West Wind Drift.
  - It is a part of the Antarctic Circumpolar Current.
  - It is a seasonal current- strong in summer and weak in winter.'
- **Kuroshio Current:** This west boundary current is also known as Japan current or Black Current. The term **"Kuroshio" in Japanese means "Black Stream ".**

- It is the Pacific analogue of the Gulf Stream in the Atlantic Ocean.
- The average surface temperature of this current is warmer than the surrounding ocean.
- This also helps in regulating the temperature of Japan, which is relatively warmer.

- **North Pacific Current:** It is formed by the collision of Kurioshio & Oyashio.
  - It circulates counterclockwise along the Western North Pacific Ocean.
- **Alaskan Current:** It results from a northward diversion of a part of the North Pacific Ocean.
- **East Australian Current:** Acts to transport tropical marine fauna to habitats in sub-tropical regions along the southeast Australian coast.
- **Florida Current:** Flows around Florida Peninsula and joins the Gulf Stream at Cape Hatteras.
- **Gulf Stream**: Western intensified current-driven mainly by wind stress.
  - It splits into North Atlantic Drift (crossing Northern Europe & southern stream) and Canary Current (recirculating of West Africa)
- **Norwegian Current:** This wedge-shaped current is one of the two dominant Arctic inflows of water.
  - It is a branch of North Atlantic Drift and sometimes also considered as an extension of the Gulf Stream.
- **Brazilian Current:** Flows along the south coast of Brazil till Rio de la Palta.
  - It joins the cold Falkland Current at the Argentine Sea making it a temperate sea.
- **Mozambique Current:** Flows between Mozambique and the island of Madagascar along the African east coast in the Mozambique Channel.
- **Agulhas Current:** Largest western boundary ocean current.
  - Flows south along the east coast of Africa.
- **Southwest Monsoon Current:** It dominates the Indian Ocean during the southwest monsoon season (June–October).
  - It is a broad eastward flowing ocean current that extends into the Arabian Sea and Bay of Bengal.

What are the Effects of Ocean Currents?

- **Climatic Conditions:** Currents influence the climatic conditions of the regions in which they flow.

- The warm Equatorial currents raise the temperature of the region in which they flow. Similarly, the cold currents lower the temperature of the places where they flow.
- For example, the British Isles would have been extremely cold without the warm North Atlantic Drift.
- The hot climate of Peru is cooled by the cold Peru Current.

◉ **Rainfall:** The winds blowing over warm currents pick up and carry moisture and bring rainfall like the North Atlantic Drift brings rainfall in some areas located along the western coasts of Europe.

- On the contrary, cold currents do not bring rainfall and make the region cooler and drier.
- The Kalahari Desert hardly experiences rainfall due to the cold Benguela current.

◉ **Fog Formation:** The meeting of the warm and the cool currents results in the formation of fog.

- The ship's face danger due to the fogs caused by the meeting of the warm currents with the cold currents.
- This has resulted in the wreckage of many ships in the past as they are not able to view icebergs due to poor visibility.

◉ **Creates Fishing Zone:** The mixing of warm and cold currents results in the deposition of planktons. Therefore, at such places, fishes can be found in abundance.

◉ **Desert formation:** Cold ocean currents have a direct effect on desert formation in west coast regions of the tropical and subtropical continents.

- There is fog and most of the areas are arid due to **desiccating effects** (loss of moisture).

◉ **Trade and Commerce:** Currents help ships to sail if they follow the directions of the currents.

- Many warm currents keep the ports of Europe ice free even during the winters. This helps in trade and commerce.

◉ **Violent Storms:** At times the meeting line of a warm and a cold current may result in a violent storm.

- The hurricanes which occur off the coast of the U.S.A. follow the line where the Gulf Stream merges with the Labrador Current.

HOW ARE OCEAN CURRENTS MEASURED?

Measurement of currents

Currents are measured by two methods they are

Direct method

Currents can be measured directly by observing the motion of water relative to the solid earth or by observing the motion of the solid earth relative to the water.

Example: A floating device tied to a weight at the ocean bottom by a wire of appropriate length depending on the depth of the current to be measured. The water speed can be measured by the speed with which a propeller is turned, and the current direction might be detected by comparing the direction of a weather wave-type pointer to a compass needle.

An example for lager device would be a float of just the right buoyancy to float at the desired depth. It should be deposited one day and then retrieved several weeks or months later and it should be attached with signal generating device to identify. The ocean current would then be known by how far and in what direction the float had gone.

Indirect method

Geostrophic technique in which the current can be measured by measuring the slope of density contours due to current flow (cf) by knowing the slope of the density gradients and knowing the latitude.

In the second method the principle of electricity and magnetism is involved. When charged particles move through a magnetic field they get deflected, opposite charges in opposite direction. Ocean currents carry many charged particles-the salt ions that are moving through the earth's magnetic field. When two metal plates are put in to this current the positive ions will go toward one plate and the negative ions towards the other. By measuring the electrical voltage thus generated between these two plates, and the orientation of the plates for maximum voltage, we can predict the speed and direction of current. All insturments dicussed so far produce information about oceanic property irrespective of the dynamic state of the ocean.

An elementary way of observing oceanic movement is the use of drifters. Drifters are platforms designed to carry instruments. But all measurements obtained from drifters are of little use unless they can be related to positions in space. A geolocation (GPS) device which transmits the drifter location to a satellite link is therefore an essential instrument on any drifter, and this turns the drifter into an instrument for the measurement of ocean currents.

Ocean currents can be measured in two ways. An instrument can record the speed and direction of the current, or it can record the east-west and north-south components of the current. Both methods require directional information. All currents meters therefore incorporate a magnetic compass to determine the orientation of the instrument with respect to magnetic north. Four classes of current meters can be distinguished based on the method used for measuring current magnitude.

Mechanical current

Meters use a propeller-type device. A Savoniur rotor or a paddle-wheel rotor to measure the current speed, and a vane to determine current direction. Propeller sensors often measures speed correctly only if they point into the current and have to be oriented to face the current all the time. Such instruments are therefore fitted with a large vane, which turns the entire instrument and with it the propeller into the current.

Propellers can be designed to have response with the angle of incidence of the flow. Two such propellers arranged at 90p will resolve current vectors and do not require an orienting vane.

The advantage of the Savonius rotor is that its rotation rate is independent of the direction of exposure to the current. A Savonius rotor current meter therefore does not have to face the current in any particular way, and its vane can rotate independently and be quite small, just large enough to follow the current direction reliably.

With the exception of the current meter that uses two propellers with cosine response set at 90p to each other, mechanical current meters measure current speed by counting propeller or rotor revolutions per unit time and current direction by determining the vane orientation at fixed intervals. In other words, these current meters combine a time integral or mean speed over a set time interval (the number of revolutions between recordings) with an instantaneous reading of current direction (the vane orientation at the time of recording). This gives only a reliable recording of the ocean current if the current changes slowly in time. Such mechanical current meters are therefore not suitable for current measurement in the oceanic surface layer where most of the oceanic movement is due to waves.

The Savonius rotor is particularly problematic in this regard. Suppose that the current meter is in a situation where the only water movement is from waves. The current that alternates back and forth, but the mean current is zero. A Savonius rotor will pick up the wave current irrespective of its direction, and the rotation count will give the impression of a strong mean current. The paddle-wheel rotor is designed to rectify this; the paddle wheel rotates back and forth with the wave current, so that its count represents the true mean current.

Mechanical current meters are robust, reliable and comparatively low in cost. They are therefore widely used where conditions are suitable, for example at depths out of reach of surface waves.

Electromagnetic current

Meters exploit the fact that an electrical conductor moving through a magnetic field induces an electrical current. Sea water is a very good conductor, and if it is moved between two electrodes the induced electrical current is proportional to the ocean current velocity between the electrodes. An electromagnetic current meter has a coil to produce a magnetic field and two sets of electrodes, set at right angle to each other, and determines the rate at which the water passes between both sets. By combining the two components the instrument determines speed and direction of the ocean current.

Acoustic current

Meters are based on the principle that sound is a compression wave that travels with the medium. Assume an arrangement with a sound transmitter, and let receiver B located downstream. If a burst of sound is generated at the transmitter it will arrive at receiver B earlier than at receiver A, having been carried by the ocean current.

A typical acoustic current meter will have two orthogonal sound paths of approximately 100mm length with a receiver/transmitter at each end. A high frequency sond pulse is transmitted simultaneously from each transducer and the difference in arrival time for the sound travelling in opposite directions gives the water velocity along the path.

Electromagnetic and acoustic current meters have no moving parts and can therefore take measurements at a very high sampling rate (up to tens of readings per second). This makes them useful not only for the measurement of ocean currents but also for wave current and turbulence measurements.

Table 1. Table of methods and devices used to measure currents.

Device	Description	
Flow Meter	Flow meters are small, often handheld devices used to measure current flow. Water current spins a propeller as it moves past the meter. The amount the propeller spins can be correlated with current speed. Some meters can also report the direction of water flow. Flow meters are useful in smaller bodies of water.	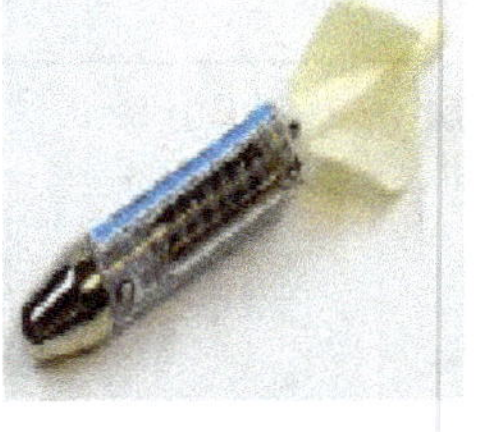

Device	Description	
Clod Cards	Clod cards are small blocks of plaster (or a similar type of material) used to measure relative flow rate between sites. As water current flows over the blocks, they dissolve. The faster the water flow, the more the clod cards dissolve. Clod cards are useful tools for measuring water flow near ocean bottom.	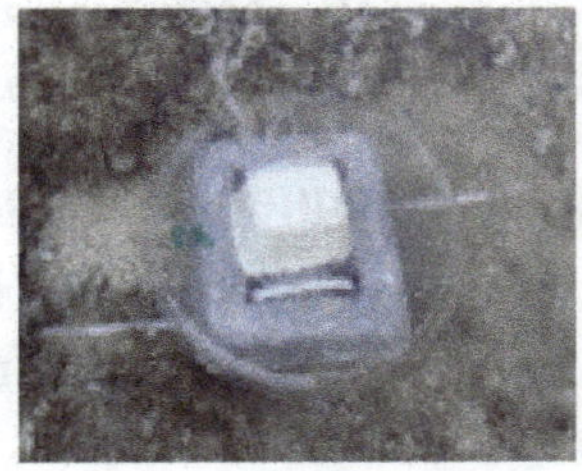
Shallow Water Drifter	Shallow water drifters float near the surface of the water and are pushed by the predominant surface current. The distance traveled, time, and direction of a drift can be measured by an observer or GPS device. The picture is a Davis drifter, which uses underwater sails moved by current flow.	
Deep Ocean Drifter	Deep ocean drifters flow with the current below the surface. They are programmed to descend to a predetermined depth for several days and then rise to the surface. While underwater, they record their position and then transmit information back to scientists when they surface. Deep ocean drifters are useful for long-term deployment in deep water as they can surface and sink through many cycles.	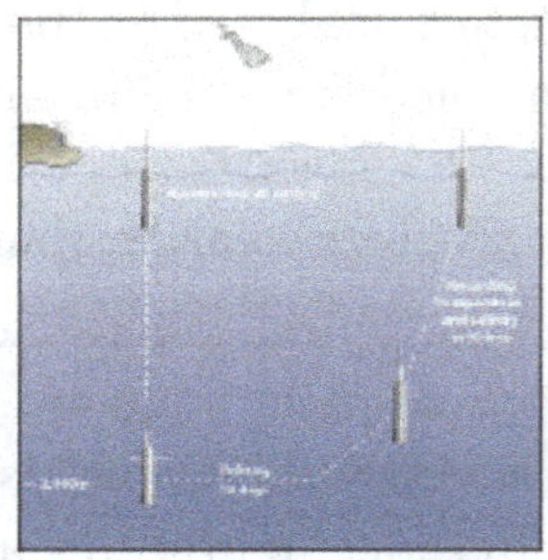
Acoustic Doppler Current Profiler (ADCP)	An ADCP emits sound pulses underwater and then measures the frequency of the sound bouncing back off of the water particles. If the water particles are moving away from the ADCP, then the frequency will be longer. If the particles are moving toward the ADCP, then the frequency will be shorter. ADCPs are useful for measuring flow in bays. They can also be mounted on the bow of ships.	
Shore Based Current Meters	Shore-based current meters send out sound signals and measure the frequency of sound bouncing back. These instruments bounce sound off wind induced surface currents. Shore-based current meters are useful when measuring surface currents. When multiple meters are used at once, current velocity maps can be generated.	

Types of Ocean Currents

Horizontal Currents

Surface Currents

They are currents that occur on the top layer of the ocean and are primarily driven by wind. They affect the water on the upper 300m of the ocean. The currents mirror the large-scale circulation of air, mostly derived from unequal heating of the planet's surface by the sun.

The currents, therefore, form rotating systems in the middle of the ocean systems, called gyres. Surface currents are responsible for redistributing heat at a planetary scale.

Deep-water Currents

As opposed to surface currents that occur on the upper surface of the ocean, deep-water currents occur deep inside the ocean. As they occur far below the surface, they are not influenced by the wind.

However, they arise as a result of variation in the density of the ocean water and are controlled by the temperature and salt content of the water.

Vertical Currents

Upwelling

Upwelling currents are currents that move from deep in the ocean heading towards the surface. They are responsible for bringing organic matter from below the ocean towards its surface.

For instance, they sweep nutrients upwards, helping some marine life. It can be seen when there are tremors or earthquakes on the surface below the ocean and the waves are pushed upwards.

In Antarctica, upwelling currents pump nitrogen and phosphates up from the deep sea to blooms of algae and other plants.

The planktons can then be eaten by crustaceans called krill, which in turn are eaten by penguins, seabirds, seals, and the baleen whales, the largest animals on earth.

Downwelling

These are currents that move material from the surface of the ocean towards its floor. Surface water can be forced downwards by the pressure of the water when currents converge or wind drives the ocean against a coastline.

It is important as the dissolved oxygen in the sediments and water below would quickly be used up by the decay of organic matter.

Additionally, anaerobic bacteria would take control of decomposition, contributing to a build-up of hydrogen sulfide and only a handful of benthic animals would survive such toxic conditions.

Causes of Ocean Currents

Solar heating

It causes water to expand. The water around the equator is roughly 8cm high than in middle latitudes. It causes a slight slope in the water and it flows down the slopes. Warm water flows towards the cold Polar Regions and cold water flows towards the warmer equatorial regions.

Wind

The Wind is responsible for ocean currents as it blows the water on the surface, causing the currents. The wind is responsible for surface currents where ocean water is redistributed based on its density and temperature.

Gravity

Gravity tends to pull items towards the surface of the earth. When the wind blows ocean water, the water piles up in the direction of the wind. Gravity, therefore, pulls the water down the 'hill' against the pressure gradient.

The salinity of the water

When water moves towards the poles, it gets cold and freezes into ice, leaving a share of salt behind. It makes the underlying water saltier, making it denser. The cold, saltier and denser water sinks to the floor of the ocean and is replaced by surface water in the process.

Temperature

Warm water tends to stay on the surface of the ocean. When the denser, saltier, and cold water moves towards the equator, where it is warmer, it warms up and becomes less dense, rising to the surface of the ocean, resulting in upwellings.

Coriolis effect

It refers to the rotation of the earth. It produces forces on all bodies moving relative to the earth. As the earth is spherical, the forces are greatly felt at the poles and least at the equator.

Coriolis effect also results in the wind changing direction, with Northern Hemisphere winds and currents moving towards the right while Southern Hemisphere currents being deflected to the left.

Underwater earthquakes

They can trigger ocean currents, moving masses of water inland. Earthquakes can also trigger downslope movements of water-saturated sediments, resulting in strong turbidity currents.

Effects of Ocean Currents

They result in rain

Warm currents lead to evaporation, which turns into the rain for the coastal areas. For instance, North Atlantic Drift brings rainfall to Western parts of Europe that results in rains throughout the year.

They can also cause deserts

Cold currents do not give rise to moisture-laden winds and, therefore, there is no rainfall in the coastal regions. The conditions cause desertification in the affected coastal regions, such as the Kalahari and Patagonia Deserts because of the cold Benguela and Falkland currents, respectively.

They can destroy marine wildlife

Some currents might be too strong and end up destroying planktons. For instance, El Nino currently destroys planktons along the Peruvian coasts. It also brings with it several diseases that kill fish.

They help with the continuity of life

In Antarctica, strong upwelling currents pump nitrogen and phosphates up from the deep sea to blooms of algae and other plants. The planktons are eaten by crustaceans called krill. The krill in turn feed penguins, seabirds, seals, and the baleen whales – the largest animals on earth.

They save up on time and costs

It goes to the advanced knowledge of currents by sailors. When currents are in their favor, they can reach their destinations in time. They can also rescue shipping time and fuel costs.

However, if they are against the sailors, they might make tore time battling the currents, and use more fuel to sail past those currents

They can result in deaths

Ocean currents can also result in deaths and destruction of property if they are strong enough to overpower ships in the ocean. Of course, to do so, they have to be very strong and probably accompanied by tremors or lightning, but in the end, they can lead to disastrous results.

Importance of Ocean Currents

Controlling the climate

Ocean currents are responsible for moving the heat from the equator and towards the poles. As such, they maintain the natural order and balance of the climate.

Critical to marine life

Marine wildlife is heavily dependent on the balance created by the ocean and is maintained by the ocean currents. The currents carry nutrients and food organisms, feeding the plants and animals that depend on them.

They also carry reproductive cells and ocean life to new places. The best example is sea turtles that lay eggs in the sand along the shores of the ocean. The ocean currents then carry the young hatchlings into the water.

They are vital for ports in the polar regions

Warm ocean currents are responsible for keeping the ports in polar regions operational. It is because they keep the ports ice-free. The North Atlantic Drift, for instance, keeps most of the European ports ice-free and operational. Norway is a prime beneficiary of this drift.

Dispersal of life forms

Ocean currents are also important as they disperse many life forms. For instance, the lifecycle of the European eel is highly sustained and influenced by ocean currents.

Transportation by humans

Humans rely on ocean currents to move some of their sea vessels, such as boats, on water. Currents are also important as they help when docking and undocking boats, speeding up shipping lanes, and keeping the ships safe, primarily in narrow waterways.

The direction of currents can also help in search and rescue missions as well as environmental disaster clean-ups.

They are responsible for some sports

Not all sports are played on dry land. Some currents create waves that can be used for competitive sports or recreational activities such as surfing. When surfing, a surfer rides with or against the wave, helping carry the surfer to shore.

HOW DO OCEAN CURRENTS AFFECT THE CLIMATE?

Maintaining the climate around the poles

Warm currents move toward the poles while the cold current move toward the equator. Such movement ensures the climate around the poles is maintained, and life around the area is not disrupted.

Causing rains

Currents are also responsible for rains. Warm currents migrate from around the equator towards the poles and at the same time, the warmth causes the water to evaporate. The same will result in rains around the coastal regions or the ocean.

Ocean water directs ocean currents

Ocean currents are continuously directed by the movement of ocean water. Forces acting on the water, such as its density, temperature, and its salinity, among others, control the currents on the ocean.

They create global conveyor belts

The global conveyor belt is a system of deep and surface currents, which move water all over the globe. The currents circulate the globe and can be instrumental in determining the climate of a region.

For instance, the North Atlantic Drift brings rainfall to Western parts of Europe, and Benguela and Falkland's currents cause desert conditions in the Kalahari and Patagonia Deserts, respectively.

CURRENTS, UPWELLING AND DOWNWELLING

The movement of surface currents also plays a role in the vertical movements of deeper water, mixing the upper water column. **Upwelling** is the process that brings deeper water to the surface, and its major significance is that it brings nutrient-rich deep water to the nutrient-deprived surface, stimulating primary production. **Downwelling** is where surface water is forced downwards, where it may deliver oxygen to deeper water. Downwelling leads to reduced productivity, as it extends the depth of the nutrient-limited layer.

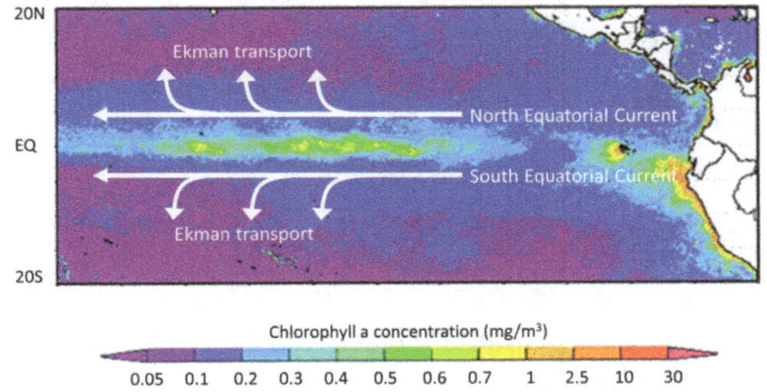

Fig. 1 Equatorial upwelling and increased productivity as a result of divergence between the north and south equatorial currents

Upwelling occurs where surface currents are diverging, or moving away from each other. As the surface waters diverge, deeper water must be brought to the surface to replace it, creating upwelling zones. The upwelled water is cold and rich in nutrients, leading to high productivity. Many of the most productive regions on Earth are found in upwelling zones. In the equatorial Pacific, the trade winds blow the North and South Equatorial Currents towards the west, while Ekman transport causes the upper layers to move to the north and south in their respective hemispheres. This creates a divergence zone, and a region of upwelling and high productivity (Fig. 1).

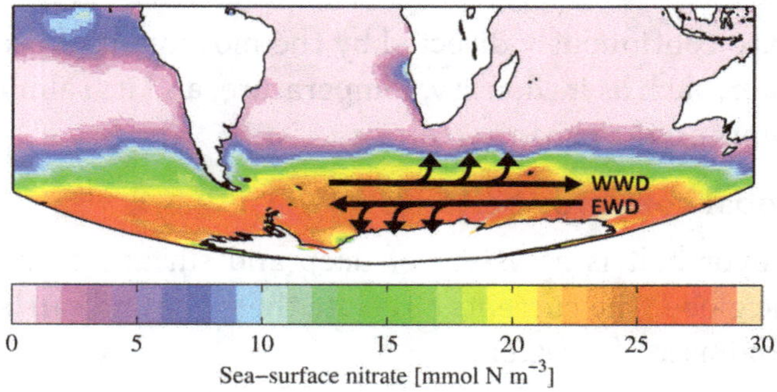

Fig. 2 High nutrient levels in the Antarctic divergence zone, as a result of the diverging West Wind Drift and East Wind Drift currents creating strong upwelling

A similar process occurs near the Antarctic continent, creating one of the most productive regions on Earth, the Antarctic divergence. In this case, the West Wind Drift (Antarctic Circumpolar Current) is flowing parallel to, but in the opposite direction of the East Wind Drift. With both currents occurring in the Southern Hemisphere, Ekman transport will be to the left, so the eastward-flowing West Wind Drift water will be transported to the north, and the westward-flowing East Wind Drift water will be transported to the south, creating a highly productive divergence zone (Fig. 2).

Downwelling occurs where surface currents converge. The converging water has nowhere to go but down, so the surface water sinks. Since surface water is usually low in nutrients, downwelling leads to low productivity zones. An example of a downwelling region is off of the Labrador coast in Canada, where the Gulf Stream, Labrador, and East Greenland Currents converge.

Coastal Upwelling

Upwelling and downwelling also occur along coasts, when winds move water towards or away from the coastline. Surface water moving away from land leads to upwelling, while downwelling occurs when surface water moves towards the land. Historically, some of the most productive commercial fishing grounds have been associated with coastal upwelling. Along the coast of California, the local prevailing winds blow towards the south. Ekman transport moves the surface layer 90° to the right of the wind, meaning the net Ekman transport is in an offshore direction. The water displaced near the coast is replaced by cold, nutrient-rich deeper water that is brought to the surface through upwelling, leading to high productivity (Fig. 3).

The same process happens off of the coast of Peru, which for a long time had the world's largest commercial fishery. Winds along the Peruvian coast blow towards the north, and since Peru is in the Southern Hemisphere, the Ekman transport is 90° to the left of the wind, which causes the surface water to move offshore and leads to upwelling and productivity. In any coastal upwelling location, if the winds reverse, surface water moves towards the shore and downwelling is the result.

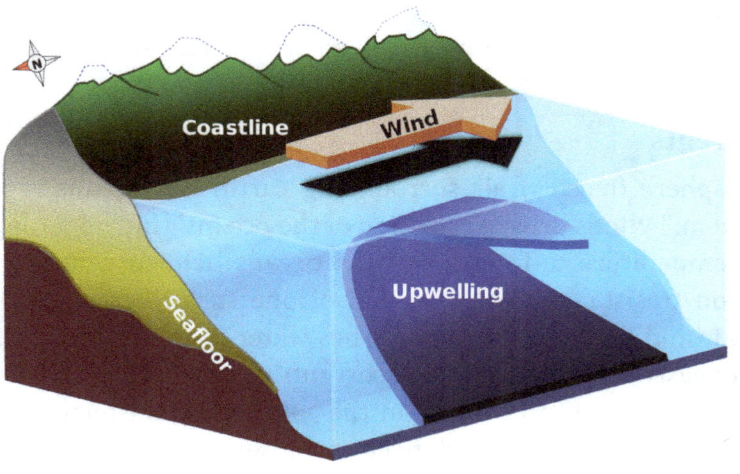

Fig. 3 Coastal upwelling. As wind blows along a coastline, Ekman transport moves the surface layer in a direction 90° to the wind. In this figure, it is to the right of the wind, indicating a Northern Hemisphere location. As surface water moves offshore it it replaced by upwelled deeper water

Upwelling can also occur due to geological features of the ocean floor. For example, as deep water currents encounter seamounts or other raised features, the water is forced upwards, bringing nutrient-rich water to the surface. This helps explain why productivity is often high in the water over seamounts.

WHAT ARE THE MAIN SURFACE CIRCULATION PATTERNS IN EACH OCEAN BASIN?

The oceans are in constant motion. Ocean currents are the horizontal and vertical circulation of ocean waters that produce a steady flow of water in a prevailing direction. Currents of ocean water distribute heat around the globe and help regulate Earth's climate, even on land. Currents carry and recycle nutrients that nourish marine (ocean) and coastal plants and animals. Human navigators depend on currents to carry their ships across the oceans. Winds drive currents of surface water. Differences in temperature and salinity (saltiness) cause water to circulate in the deep ocean. The rotation of the Earth, the shape of the seafloor, and the shapes of coastlines also determine the complex pattern of surface and deep ocean currents.

Ocean water is layered. The shallowest water, called surface water, is warmer, fresher, and lighter than deep water, which is colder, saltier, and denser. The boundary between surface and deep water is a thin layer marked by an abrupt change of temperature and salinity. This layer, called the thermocline, exists in most places in the oceans. Surface and deep water only mix in regions where specific conditions allow deep water to rise or surface water to sink. Many organisms swim freely across the thermocline, and the remains of plants and animals continuously rain down through the deep water to the seafloor. However, most organisms live, or at least feed, close to the ocean surface where microscopic plants called phytoplankton float freely and absorb the sunlight they need to live. Very little light penetrates the surface water.

Surface currents

Earth's atmosphere (mass of air surrounding Earth) and oceans together form a "coupled system." Winds drive circulation of the oceans' thin upper layer of surface water, and temperature differences in the oceans help to generate atmospheric winds. Friction (resistance to the motion of one surface over another) between the moving air and the water surface pushes water in the direction of the blowing winds. Earth's eastward rotation causes currents to deflect (bend) to the right in the northern hemisphere and to the left in the southern hemisphere, a phenomenon called the Coriolis effect. Coriolis deflection causes clockwise circulation of wind-driven surface ocean currents in the northern hemisphere and counterclockwise circulation in the southern hemisphere.

Warm surface currents

The subtropical trade winds, or trades, are strong, steady winds that blow warm water from west to east on either side of the equator (an imaginary line around Earth halfway between the North and South Poles), thereby creating west-flowing equatorial currents in the major oceans. The trades and equatorial currents push piles of warm water into the western halves of the Pacific, Indian, and Atlantic Oceans. Water flows down and away from the centers of the mounds, not unlike pancake batter spreading out on a griddle. The trades continue to push the water to the west, and Coriolis deflection guides it northward in the northern hemisphere and southward in the southern hemisphere. Warm, fast currents, called western boundary currents, flow away from the tropical warm pools toward the poles in the western halves of the ocean basins.

The Gulf Stream is the western boundary current in the North Atlantic. It flows north along the southeastern coast of the United States and then crosses the Atlantic on a diagonal path. Warm Gulf Stream waters create unusually mild climates in northern locations. Gulf Stream waters keep Bermuda balmy, Ireland green, and England foggy. The Gulf Stream is the major shipping route from North America to

Europe. The western boundary current in the North Pacific, called Kuroshio, likewise warms the islands of Japan and carries ships toward the Pacific Northwest. Western boundary currents in the southern hemisphere, the Brazil Current in the South Atlantic, East Australian Current in the South Pacific, and Aguellas Current in the Indian Ocean flow south from the equator.

THE CORIOLIS EFFECT

The paths of objects traveling on or above the Earth's surface are deflected to the right in the northern hemisphere and to the left in the southern hemisphere. This phenomenon, called the Coriolis effect, is named for French mathematician Gaspard-Gustave de Coriolis (1792–1843), who first explained it in 1835. Ocean gyres rotate clockwise in the northern hemisphere and counterclockwise in the southern hemisphere because of the Coriolis effect.

The Coriolis effect is sometimes called the Coriolis force. This is incorrect because nothing forces objects' paths to curve. The moving Earth is the frame of reference used to measure the route of a traveling object. A point on the ground at the equator has to move much faster to make its daily revolution than a point near the North or South Pole. Imagine an old phonograph record with a piece of graph paper glued to its surface. If you use a pencil to draw a straight line from the center of the turning record to its edge, the line drawn is a curve even though your pencil moved in a straight line. By the same token, the path of a missile fired straight from the North Pole toward the equator will curve to the east. Proper calculation of the Coriolis effect is very important for air traffic controllers and long-range missile programmers.

The Coriolis effect only creates a noticeable deflection for objects traveling long distances along north-south paths. The common myth that whirlpools and toilet flushes rotate the opposite directions across the equator is incorrect. A toilet is too small for the Coriolis effect to change the path of water in the bowl.

Cool surface currents

Cold surface currents carry cool water from the poles toward the equator along the west coasts of the continents. The cool eastern boundary currents are generally shallower and weaker than the western boundary currents. Cold water flowing from the Arctic Ocean at the North Pole feeds the eastern boundary currents in the northern hemisphere, namely the California Current in the Pacific and the Canary Current in the Atlantic. The Antarctic Circumpolar Current (ACC) that encircles the ice-covered Antarctic continent supplies the cool water to the eastern boundary currents of the southern hemisphere—the Peru, Benguela, and West Antarctic Currents. The ACC is an exception to the general rule that surface currents are shallow. It extends from the sea surface to the seafloor in several places.

Gyres

Surface water circulates in oceans in massive circular patterns called gyres. The major surface currents (eastern boundary, western boundary, and equatorial current) in each ocean link to form a circle. Gyres are clockwise in the northern hemisphere and counterclockwise in the southern hemisphere. For example, a rubber duck dropped into the ocean near San Diego might float south on the California Current to the North Equatorial Current, west across the Pacific Ocean to the Kuroshio western boundary current, and then back across the northern Pacific to British Columbia. In a few years, the California Current might return the duck to San Diego. Ocean researches have actually conducted many such experiments. One important study tracked 29,000 plastic bathtub toys that spilled from a cargo ship in the North Pacific.

The infamous "triangle trade" between Europe, Africa, and North America in the eighteenth and nineteenth centuries relied on the North Atlantic Gyre. European slave ships arrived in Africa via the Canary Current, then carried slaves to the sugar plantations of the Caribbean on the North Equatorial Current. Having left off slaves and picked up sugar, they rode the Gulf Stream north to the rum distilleries of New England and the liquor shops in Europe. Hurricanes that form in the tropical Atlantic ride the North Equatorial Current toward the Caribbean Islands and then follow the Gulf Stream toward the southeast coast of the United States.

DEEP OCEAN CURRENTS

Deep ocean currents are driven by differences in temperature and salinity. They are generally unaffected by surface currents. Deep water is colder, saltier, and denser than surface water. Deep water forms in polar regions where warmer surface water cools and sinks beneath the Arctic ice cap (permanent ice covering) or Antarctic ice shelves (permanent ice large enough to cover most of a land mass). Salinity increases near the ice caps because seawater forms freshwater ice when it freezes. The salt stays behind and the remaining liquid water becomes saltier. A "global conveyor belt" carries deep water south through the Atlantic, around Antarctica, and north into the Pacific, Indian, and Atlantic Oceans. It could take the molecules in a drop of water more than a thousand years to make a complete circuit of this global deep ocean current.

HOW DO SEA ICE AND ICEBERGS FORM?

Ice exists in the natural environment in many forms. Ice sheets are the largest forms of glaciers in the world and have smaller outlet glaciers or ice streams near their margins. In some places where the ice sheets reach the ocean, floating glacier tongues are formed. Icebergs are floating ice masses that have broken away from ice shelves, glacier tongues, or directly from the grounded ice sheet in some locations. Sea ice, which is produced when saline ocean water is cooled below its freezing temperature of approximately -2°C or 29°F, extends on a seasonal basis over great areas of the ocean.

Sea ice and icebergs are both carried by winds and currents into warmer waters. Melt water from sea ice, ice shelves, glacier tongues, and icebergs does not contribute to sea level rise, because these ice masses already displace an equivalent amount of sea water. However, sea level rise is caused by the flow of grounded glacial ice into the ocean and by surface or subsurface melt water discharged from the glacier, if the sum of those amounts exceeds the amount of ice accumulated from snowfall on the glacier or ice sheet.

Types of Ice

The ice that is seen floating on the ocean's surface comes from one of two sources. Glacial ice is formed from the accumulation and compression of snow into glaciers, that then break apart and release ice to the ocean. Because glaciers can be several kilometers thick the icebergs that break off of them can be very large; so the tall icebergs at sea always come from glacial ice sheets. Sea ice refers to the ice formed from the freezing of sea water, and rarely exceeds a thickness of several meters. Sea ice covers about 7% of the ocean at any time, and makes up about 66% of the Earth's permanent ice cover by area, but only 0.1% of the ice in terms of volume. This is because sea ice is a vast but thin sheet of cover compared to the glacial ice caps that are more localized but may be several kilometers thick.

Sea ice cover around Antarctica fluctuates between about 21 million km^2 in winter to around 1.3 million km^2 in the summer, with most Antarctic sea ice lasting only a year. Seasonal changes in ice cover are less pronounced in the Arctic, from about 14 million km^2 in winter to 6.5 million km^2 in summer. About half of the sea ice in the Arctic lasts more than a year to become multi-year ice. This difference arises because Antarctica is surrounded by water, so the ice expands into warmer water and eventually melts. The Arctic Ocean is enclosed by continents, so only about %10 of the ice escapes into the Atlantic between Greenland and Spitzbergen. The rest is trapped and becomes multi-year ice or perennial ice, averaging around 7 years old, and 3-5 m thick, compared to first year ice at 1-2 m thick.

Sea ice formation

Because of the salt content, seawater begins to freeze at about -1.8° C, a lower temperature than for fresh water. Ice formation begins at the surface with the formation of small needle-like ice crystals called **frazil**, which accumulate and make the water appear slushy and cloudy; this stage is referred to as **grease ice** (Fig. 4A). In calmer water these small crystals can freeze together into a thin surface layer called **nilas**, which can reach a thickness of up to 10 cm (Fig. 4B).

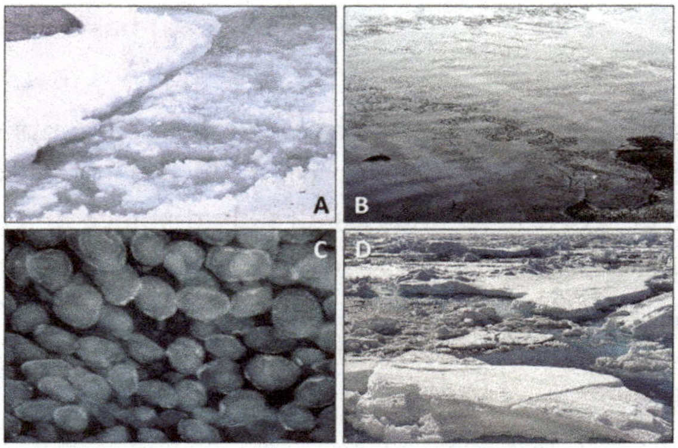

Fig. 4 Stages in sea ice formation. A) grease ice, B) nilas, C) pancake ice, D) ice floes. (A) National Park Service [Public domain], via Wikimedia Commons; B) Brocken Inaglory

Wave action can break up the nilas into small mats 1-2 m across, which then bump into each other and form rounded shapes with raised edges, called pancake ice (Fig. 4C). If temperatures remain cold the pancake ice freezes together into solid ice floes, a hard surface covering the ocean (Fig. 4 D). Ice floes then freeze together into ice fields.

Once ice floes form, the water underneath becomes insulated and heat loss to the atmosphere declines, so the water no longer cools and no more ice formation occurs. As a result, young sea ice is usually relatively thin, not more than 3-4 m thick. The ice can get thicker through precipitation; there is not a lot of precipitation at the poles, but due to the low temperatures, whatever does occur tends to accumulate rather than melt. Over time the accumulated ice and snow can add to the overall thickness of the sea ice, but it will still never approach the thickness of glacial ice.

As sea ice crystals form, most of the salt is excluded, so sea ice contains much less salt than seawater, and can be melted for drinking if needed. But about 20% of the salt remains trapped in pockets of water between the ice crystals. As ice forms and salts are excluded into these pockets, the salinity of the remaining water increases and it can become too salty to freeze. These unfrozen pockets of briny water make sea ice a little softer and more slushy than fresh water ice, which is harder and more rigid. Eventually most of this brine leaks out, and the sea ice becomes more solid, but when it is "young ice" it can be more dangerous to walk on than fresh water ice of the same thickness. For example, 7-8 cm of fresh water ice is enough to support the weight of a single person, but you would need at least 15 cm of sea ice to do the same.

The very cold, dense briny water leaks out of the ice and sinks. The brine is "supercooled"; it is cooled below the normal freezing point of seawater, but remains liquid due to the high salt content. When this supercooled brine comes into contact with the surrounding water, it causes the water around it to freeze, creating hollow ice stalactites, or "brinicles" that can be several meters long. Brine continues to flow through the hollow brinicle, and the brinicle grows downwards (see below for an amazing time-lapse video of brinicle formation).

When the large sheets of sea ice are formed, they exist in one of two formations. **Fast ice**, or **land-fast ice**, refers to the large, solid ice sheets that are attached to land. The **pack ice** consists of the smaller, free-floating pieces of sea ice. They may have formed independently, or may have broken off from the fast ice.

The floating pack ice dampens waves and currents, buffering the sea surface from motion. So changes in the distribution of the pack ice can lead to changes in current patterns, and even alter ecosystem structure. But the pack ice is also subject to the currents flowing underneath, and the ice sheets are constantly on the move, breaking up or being pushed together. When pieces of ice converge, they often buckle and crack, or override one another as in convergent lithospheric plate boundaries. These collisions can create tall, jagged pressure ridges, that may extend for several kilometers, and that create hazards for polar explorers navigating the ice.

In the polar oceans, ice cover is not uniform. There are a number of areas where there is consistently open water, even though the areas surrounding them are covered by ice. These regions of persistent open water are called **polynyas** (Fig. 5). Polynyas may be the result of currents or winds moving the ice, or areas of warmer water that prevent ice formation.

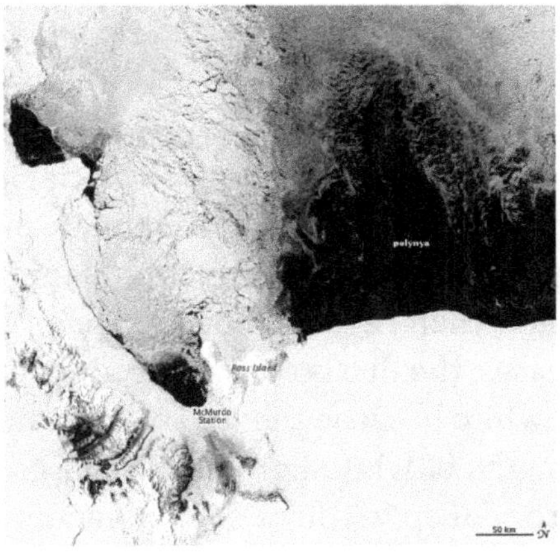

Fig. 5 A polynya near McMurdo Station in Antarctica

Iceberg

Icebergs are large chunks of ice that break off from glaciers. This process is called calving. Icebergs float in the ocean, but are made of frozen freshwater, not saltwater.

Most icebergs in the Northern Hemisphere break off from glaciers in Greenland. Sometimes they drift south with currents into the North Atlantic Ocean. Icebergs also calve from glaciers in Alaska.

In the Southern Hemisphere, almost all icebergs calve from the continent of Antarctica.

Some icebergs are small. Bergy bits are floating sea ice that stretch no more than 5 meters (16.5 feet) above the ocean. Growlers are even smaller.

Icebergs can also be huge. Some icebergs near Antarctica can be as big as Sicily, the largest island in the Mediterranean Sea. As little as one-eighth of an iceberg is visible above the water. Most of the mass of an iceberg lies below the surface of the water. This is where the phrase "tip of the iceberg" came from, meaning only part of an idea or problem is known.

Many of the most spectacular ice formations in the ocean take the form of icebergs. Remember that the largest icebergs are not made of sea ice; they are floating pieces of glaciers that have broken off, or **calved** from the glacier tongue, and thus they are composed of fresh water ice.

Fig. 6 Left: Changes in ice crystal structure and density as they are compacted from snow into glacial ice

Glacial ice forms from layers of snow that accumulate over time. The weight of the accumulated layers compresses the snow into a granular form known as **firn** that has a higher density and less air than regular snow. As the pressure from the continued snow accumulation increases, the firn is compressed into even denser glacial ice (Fig. 6). Most icebergs appear white because the ice still contains a lot of air bubbles that scatter all of the wavelengths of white light. But icebergs composed of older ice, or highly compressed ice from deep within a glacier can appear a deep shade of blue (Fig. 6). This ice contains much less air, and larger, denser ice crystals that absorb the longer (red) wavelengths of light, and transmit and scatter the shorter, blue wavelengths. The longer the path that light travels through the ice, the more the long wavelengths are absorbed, and the bluer the ice appears.

Icebergs come in a wide range of sizes, from a few meters across to hundreds of kilometers long. Table 2 shows the size categories used in the North Atlantic (in the Antarctic a different scale is used, as the icebergs tend to be larger there).

Table 2. Iceberg size classifications for the North Atlantic

Size Category	Height (m)	Length (m)
Growler	< 1	< 5
Bergy Bit	1-5	5-15
Small	5-15	15-60
Medium	16-45	61-120
Large	46-75	121-200
Very Large	> 75	> 200

Icebergs can also be classified based on their shapes (Fig. 7). The primary distinction is between tabular and non-tabular icebergs. Tabular icebergs have steep sides and a flat top, and the length is greater than five times the height. Non-tabular icebergs include any icebergs that are not tabular, and these can be further subdivided into a few categories. Domed icebergs have a rounded top, pinnacled icebergs have tall spires, and wedge icebergs have a steep face next to a more gradually sloping side. Drydock icebergs have a water-covered channel running through it, potentially large enough for boats to pass through; hence the name. Finally, blocky icebergs have a flat top and steep sides, but their length to height ratio is not as great as it is for a tabular iceberg. Regardless of the shape of an iceberg, what we see above the surface only represents about 10% of the total mass, with the rest of the ice remaining submerged.

Fig. 7 Icebergs can be classified based on their shapes. Clockwise from top left: tabular, domed, pinnacled, blocky, drydock, and wedge icebergs.

Icebergs in the Arctic tend to be smaller and more non-tabular than Antarctic icebergs, because most of them formed as irregular chunks of glaciers that calved off into the ocean, primarily around Greenland. It is estimated that the glaciers around Greenland and the Canadian Arctic calve off 300 billion cubic meters of icebergs each year. Because Antarctica has much larger ice sheets, the icebergs produced in the Southern Hemisphere are usually larger than those in the Arctic, and they tend to be more tabular in shape, as entire sections of the ice sheet break off at once. The largest iceberg ever recorded, the B-15 iceberg, broke off of the Antarctic ice sheet in 2000. It measured 295 km long and 37 km wide, giving it an area of about 11,000 square kilometers; about the size of Connecticut. Over the next few years B-15 drifted along the Antarctic coast before being broken up into smaller pieces (Fig. 8).

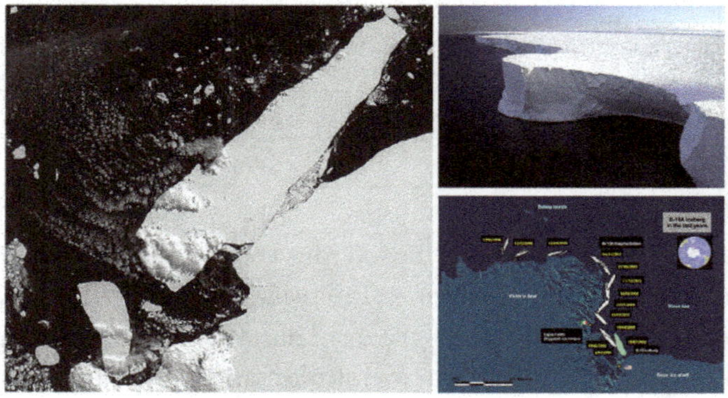

Fig. 8 Left: Iceberg B-15 soon after breaking off of the Antarctic ice sheet in 2000

All of these icebergs floating through the polar oceans are potentially hazardous to ship traffic, as we all know from the tragic story of the Titanic. In response to that disaster, the North Atlantic maritime nations established the International Ice Observation and Ice Patrol Service in 1914, to monitor icebergs threatening the shipping lanes. Interestingly, the original charge of the patrol also included searching for and destroying derelict ship hulls that were still floating around, as these were also a potential hazard. While the patrols were successful at destroying these abandoned ship hulls, they achieved far less success at attempting to destroy the icebergs. Eventually they gave up on iceberg destruction and focused their efforts on monitoring iceberg movements. Today, the service still operates as the International Ice Patrol, operated by the U.S. Coast Guard. Twice each day the patrol issues alerts with the position and potential tracks of existing icebergs, and the extent of the ice cover, reporting on the roughly 600 icebergs per year that intrude below 48° N latitude, the northern limit of the major shipping lines. No such iceberg patrol exists around Antarctica, as there is much less shipping traffic at those latitudes.

HOW MUCH OF AN ICEBERG IS VISIBLE ABOVE THE SURFACE?

Archimedes' Principle states that the upwards buoyant force of an object in water is equal to the weight of the water displaced by the object. We know that the density of an iceberg (r_i) is around 0.917 g/cm³. The weight (w_i) of the iceberg is equal to its mass (m_i) x the acceleration due to gravity (g), which is 9.8 m/s². Since density = mass/volume (V), the weight of the iceberg is:

$$w_i = r_i V_i g$$

Since the iceberg is floating at equilibrium, the weight of the iceberg (w_i) is equal to the weight of the water it displaces (w_w). The weight of water displaced is therefore:

$W_w = r_w V_w g$

where the density of seawater (r_w) is about 1.024 g/cm³. Now we have:

$W_i = W_w$

$r_i V_i g = r_w V_w g$

so $V_w = (r_i/r_w)V_i = (1.024/0.917)V_i = 0.89\ V_i$.

In other words, the volume of water displaced (V_w) is equal to about 89% of the volume of the iceberg (V_i). This means that 89% of the iceberg is submerged, leaving around 11% of the ice exposed above the surface.

DEEP OCEAN CURRENTS

Deep ocean currents in oceans are caused by a large amount of sinking surface water. Surface water is the upper layer of water closest to the top surface. The sun can easily reach this top layer, heat up the surface water and evaporate some of the water. When the surface water becomes extremely cold, the lower temperature and the additional salt causes the surface water to become more dense than the water beneath it, and thus the surface water sinks down to the deep water layers of the ocean in a circulation process known as thermohaline circulation. Thermohaline circulation, or the sinking of highly dense surface water, is the source of deep currents in oceans.

Where They Occur

Thermohaline circulation can only develop in extremely cold regions where the temperature of the air is low enough to make the surface water very cold, highly salty and more dense than the water beneath it. Thus, deep currents generally occur in the higher latitude regions of the Earth, such as North Atlantic Deep Water and Antarctic Bottom Water, and from these frigid poleward regions the deep currents flow at a relatively slow pace towards the equator.

Characteristics

After the thermohaline circulation process, the surface water that sinks down to the deep ocean does not mix well with the water beneath it, and thus it is easy to identify the sinking water masses using scientific data. Deep currents can be distinguished by the extremely cold water temperatures, the relatively high concentration of oxygen and the high salt levels that all result from sinking surface water.

Circulation Pattern

Many deep currents follow a specific circulation pattern as they travel around the planet, and the pattern usually forms a cycle. Most sinking deep water currents form in the North Atlantic, near Iceland, and from there the deep current begins its circulation pattern. The highly dense water in the deep current flows southward passed the southern edge of Africa, travels across the southern Indian Ocean, flows passed the east side of Australia, and merges into the North Pacific. Once the deep current enters the North Pacific, increasing temperatures cause a lower density in the deep water, and in turn the water becomes more buoyant and rises up to the surface again.

The surface water in the North Pacific then flows south, sliding between Asia and Australia, wraps around the southern edge of Africa again—but this time moving west—and then flows across the South Atlantic. From the South Atlantic, the water connects with the Gulf Stream and flows up north again. Once it returns to the colder, higher latitudes of the North Atlantic, the dense surface water sinks back down to the lower deep water, forms a deep current and repeats the entire cycle again.

CHAPTER

WAVES AND WATER DYNAMICS 8

Waves come in many shapes and sizes; a 100 foot wave might be a surfer's dream, but a ship captain's nightmare. What was the largest wave ever recorded? 50 feet? 100 feet? Not even close. That record belongs to a wave created in Lituya Bay, Alaska, on July 9, 1958. On that day, a magnitude 7.8 earthquake caused a massive rockslide that slid down a mountainside and into the headwaters of the bay. The rockslide created a splash wave that was high enough to flatten vegetation up to 1722 ft (525 m) above sea level! The wave then moved through the narrow bay towards the sea, destroying a number of fishing boats along the way. Miraculously, a father and son on one fishing boat were carried above the trees by the wave, and survived to tell the story. This is by far the largest wave, a megatsunami, ever reliably recorded. The waves we will discuss in this chapter may not be quite that dramatic, but it is still important to know how they form, how they are propagated, and what happens to them as they interact with the shore.

WAVE BASICS

Waves generally begin as a disturbance of some kind, and the energy of that disturbance gets propagated in the form of waves. We are most familiar with the kind of waves that break on shore, or rock a boat at sea, but there are many other types of waves that are important to oceanography:

- **Internal waves** form at the boundaries of water masses of different densities (i.e. at a pycnocline), and propagate at depth. These generally move more slowly than surface waves, and can be much larger, with heights exceeding 100 m. However, the height of the deep wave would be unnoticeable at the surface.
- **Tidal waves** are due to the movement of the tides. What we think of as tides are basically enormously long waves with a wavelength that may span half the globe. Tidal waves are not related to tsunamis, so don't confuse the two.

- **Tsunamis** are large waves created as a result of earthquakes or other seismic disturbances. They are also called seismic sea waves.

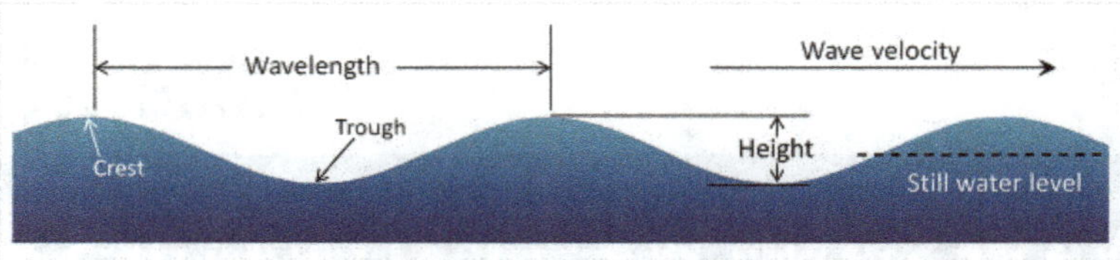

Fig 1. Components of a basic wave

- **Splash waves** are formed when something falls into the ocean and creates a splash. The giant wave in Lituya Bay that was described in the introduction to this chapter was a splash wave.
- **Atmospheric waves** form in the sky at the boundary between air masses of different densities. These often create ripple effects in the clouds.

There are several components to a basic wave (Fig. 1):

- **Still water level:** where the water surface would be if there were no waves present and the sea was completely calm.
- **Crest:** the highest point of the wave.
- **Trough:** the lowest point of the wave.
- **Wave height:** the distance between the crest and the trough.
- **Wavelength:** the distance between two identical points on successive waves, for example crest to crest, or trough to trough.
- **Wave steepness:** the ratio of wave height to length (H/L). If this ratio exceeds 1/7 (i.e. height exceeds 1/7 of the wavelength) the wave gets too steep, and will break.

There are also a number of terms used to describe wave motion:

- **Period:** the time it takes for two successive crests to pass a given point.
- **Frequency:** the number of waves passing a point in a given amount of time, usually expressed as waves per second. This is the inverse of the period.
- **Speed**: how fast the wave travels, or the distance traveled per unit of time. This is also called celerity (c), where c = wavelength × frequency

Therefore, the longer the wavelength, the faster the wave.

Although waves can travel over great distances, the water itself shows little horizontal movement; it is the *energy* of the wave that is being transmitted, not the water. Instead, the water particles move in circular orbits, with the size of the orbit equal to the wave height. This orbital motion occurs because water waves contain components of both longitudinal (side to side) and transverse (up and down) waves, leading to circular motion. As a wave passes, water moves forwards and up over the wave crests, then down and backwards into the troughs, so there is little horizontal movement. This is evident if you have ever watched an object such as a seabird floating at the surface. The bird bobs up and down as the wave pass underneath it; it does not get carried horizontally by a single wave crest.

The circular orbital motion declines with depth as the wave has less impact on deeper water and the diameter of the circles is reduced. Eventually at some depth there is no more circular movement and the water is unaffected by surface wave action. This depth is the **wave base** and is equivalent to half of the wavelength (Fig. 2). Since most ocean waves have wavelengths of less than a few hundred meters, most of the deeper ocean is unaffected by surface waves, so even in the strongest storms marine life or submarines can avoid heavy waves by submerging below the wave base.

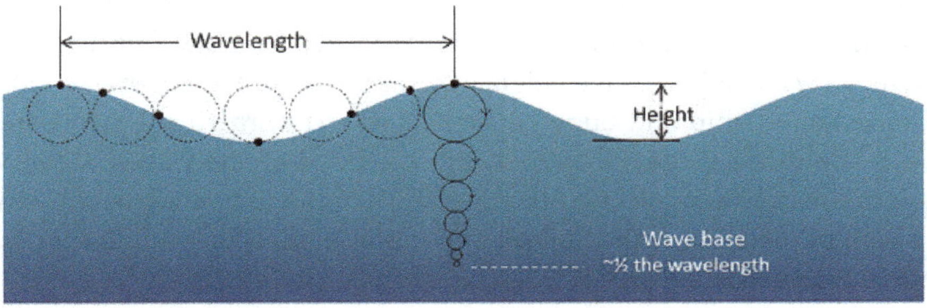

Fig. 2. Orbital motion of water within a wave, extending down to the wave base at a depth of half of the wavelength

Waves at Sea

Most ocean waves are generated by wind. Wind blowing across the water's surface creates little disturbances called **capillary waves**, or ripples that start from gentle breezes. Capillary waves have a rounded crestwith a V-shaped trough, and wavelengths less than 1.7 cm. These small ripples give the wind something to "grip" onto to generate larger waves when the wind energy increases, and once the wavelength exceeds 1.7 cm the wave transitions from a capillary wave to a wind wave. As waves are produced, they are opposed by a restoring force that attempts to return the water to its calm, equilibrium condition. The restoring force of the small capillary waves is surface tension, but for larger wind-generated waves gravity becomes the restoring force.

As the energy of the wind increases, so does the size, length and speed of the resulting waves. There are three important factors determining how much energy is transferred from wind to waves, and thus how large the waves will get:

- Wind **speed.**
- The **duration** of the wind, or how long the wind blows continuously over the water.
- The distance over which the wind blows across the water in the same direction, also known as the **fetch.**

Increasing any of these factors increases the energy of wind waves, and therefore their size and speed. But there is an upper limit to how large wind-generated waves can get. As wind energy increases, the waves receive more energy and they get both larger and steeper (recall from section 10.1 that wave steepness = height/wavelength). When the wave height exceeds 1/7 of the wavelength, the wave becomes unstable and collapses, forming whitecaps.

The ocean surface represents an irregular mixture of hundreds of waves of different speeds and sizes, all coming from different directions and interacting with each other. A histogram of wave heights within this mixture reveals a bell-shaped curve (Fig. 3). In addition to basic statistics such as mode (most probable), median and mean wave height, wave heights are also reported in other ways. Marine weather forecasts and ship and buoy data often report **significant wave height (H_s)**, which is the mean height of the largest one-third of the waves. Mean wave height is approximately equal to two-thirds of the significant wave height. Finally, there is the minimum height of the highest 10% of waves (the 90th percentile of wave heights), often expressed as $H_{1/10}$.

Under strong wind conditions, the ocean surface becomes a chaotic mixture of choppy, whitecapped wind-generated waves. The term **sea state** describes the size and extent of the wind-generated waves in a particular area. When the waves are at their maximum size for the existing wind speed, duration, and fetch, it is referred to as a fully developed sea. The sea state is often reported on the **Beaufort scale,** ranging from 0-12, where 0 means calm, windless and waveless conditions, while Beaufort 12 is a hurrican.

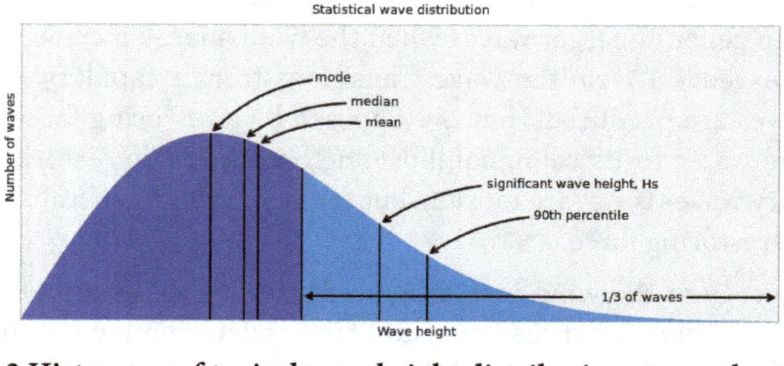

Fig. 3 Histogram of typical wave height distribution at sea, showing common statistical measurements

A fully developed sea often occurs under stormy conditions, where high winds create a chaotic, random pattern of waves and whitecaps of varying sizes. The waves will propagate outwards from the center of the storm, powered by the strong winds. However, as the storm subsides and the winds weaken, these irregular seas will sort themselves out into more ordered patterns. Recall that open ocean waves will usually be deep water waves, and their speed will depend on their wavelength. As the waves move away from the storm center, they sort themselves out based on speed, with longer wavelength waves traveling faster than shorter wavelength waves. This means that eventually all of the waves in a particular area will be traveling with the same wavelength, creating regular, long period waves called **swell.** We experience swell as the slow up and down or rocking motion we feel on a boat, or with the regular arrival of waves on shore. Swell can travel very long distances without losing much energy, so we can observe large swells arriving at the shore even where there is no local wind; the waves were produced by a storm far offshore, and were sorted into swell as they traveled towards the coast.

Because swell travels such long distances, eventually swells coming from different directions will run into each other, and when they do they create interference patterns. The interference pattern is created by adding the features of the waves together, and the type of interference that is created depends on how the waves interact with each other (Fig. 4).

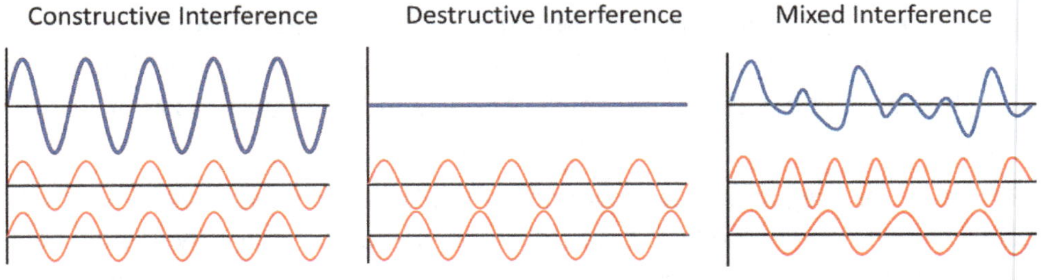

Fig. 4 Wave interference patterns. In constructive interference the source waves (red) are completely in phase, and when added together produce waves that are larger than the original waves (blue). In destructive interference the source waves are out of phase, so they cancel each other out and produce waves that are smaller than the originals. In mixed interference, constructive and destructive interference occur at various point, creating an irregular wave pattern

Constructive interference occurs when the two waves are completely in phase; the crest of one wave lines up exactly with the crest of the other wave, as do the troughs of the two waves. Adding the two crest together creates a crest that is higher than in either of the source waves, and adding the troughs creates a deeper trough than in the original waves. The result of constructive interference is therefore to create waves that are larger than the original source waves.

In **destructive interference**, the waves interact completely out of phase, where the crest of one wave aligns with the trough of the other wave. In this case, the crest and the trough work to cancel each other out, creating a wave that is smaller than either of the source waves. In reality, it is rare to find perfect constructive or destructive interference as displayed in Fig. 4. Most interference by swells at sea is **mixed interference**, which contains a mix of both constructive and destructive interference. The interacting swells do not have the same wavelength, so some points show constructive interference, and some points show destructive interference, to varying degrees. This results in an irregular pattern of both small and large waves, called **surf beat**.

It is important to point out that these interference patterns are only temporary disturbances, and do not affect the properties of the source waves. Moving swells interact and create interference where they meet, but each wave continues on unaffected after the swells pass each other.

About half of the waves in the open sea are less than 2 m high, and only 10-15% exceed 6 m. But the ocean can produce some extremely large waves. The largest wind wave reliably measured at sea occurred in the Pacific Ocean in 1935, and was measured by the navy tanker the USS Ramapo. Its crew measured a wave of 34 m or about 112 ft high! Occasionally constructive interference will produce waves that are exceptionally large, even when all of the surrounding waves are of normal height. These random, large waves are called **rogue waves**. A rogue wave is usually defined as a wave that is at least twice the size of the significant wave height, which is the average height of the highest one-third of waves in the region. It is not uncommon for rogue waves to reach heights of 20 m or more.

Rogue waves are particularly common off of the southeast coast of South Africa, a region referred to as the "wild coast." Here, Antarctic storm waves move north into the oncoming Agulhas Current, and the wave energy gets focused over a narrow area, leading to constructive interference. This area may be responsible for sinking more ships than anywhere else on Earth. On average about 100 ships are lost every year across the globe, and many of these losses are probably due to rogue waves.

Waves in the Southern Ocean are generally fairly large because of the strong winds and the lack of landmasses, which provide the winds with a very long fetch, allowing them to blow unimpeded over the ocean for very long distances. These latitudes have been termed the "Roaring Forties", "Furious Fifties", and "Screaming Sixties" due to the high winds.

TSUNAMI WARNING SYSTEM

Most tsunamis are caused by large earthquakes below or near the ocean floor, but can also be caused by landslides, volcanic activity, certain types of weather, and near-earth objects. In the deep ocean, tsunami waves are often barely noticeable, but can move as fast as a jet plane, over 500 mph. As they enter shallow water near land, they slow down and grow in height, and currents intensify.

Most tsunami waves are less than 10 feet high, but can exceed 100 feet in extreme events. When a tsunami comes ashore, it may look like a fast-rising flood or a wall of water. Sometimes, before the water rushes on land, it will drain away suddenly, showing the ocean floor like a very low, low tide. A large tsunami can flood low-lying coastal areas more than a mile inland. The first wave may not be the largest or most damaging. Waves may repeatedly flood and recede from the land for many hours.

A tsunami can be very dangerous to life and property on the coast. It can produce unusually strong currents, rapidly flood the land, and cause great destruction. The flow and force of the water and debris it carries can destroy boats, vehicles, and buildings; cause injuries; and take lives as the tsunami moves across land and returns to the sea. Dangerous currents and flooding may last for days. Even small tsunamis can be dangerous. Strong currents can injure and drown swimmers and damage or destroy boats in harbours.

Tsunami detection and observation

The NOAA (National Oceanic and Atmospheric Administration) Tsunami Program runs the U.S. Tsunami Warning System. This includes monitoring for tsunamis and the earthquakes that cause them to provide timely and accurate tsunami messages. NOAA's success in fulfilling this important mission relies on the ability to quickly detect a tsunami, which is accomplished through networks of advanced observation systems.

NOAA operates two tsunami warning centers, which are staffed 24 hours a day, 7 days a week. The National Tsunami Warning Center serves the continental United States, Alaska, and Canada. The Pacific Tsunami Warning Center directly serves the Hawaiian Islands, the U.S. Pacific and Caribbean territories, and the British Virgin Islands and is the primary international forecast center for the Pacific and Caribbean.

When an earthquake occurs, seismic networks provide information about its location, depth, and magnitude to help the warning centers determine if it may have generated a tsunami and if they should issue tsunami messages. The U.S. Geological Survey and its partners operate the primary U.S. seismic networks.

If the earthquake meets certain criteria, the warning centers use sea-level data to determine the existence of a tsunami and refine or cancel messages. NOAA has established and maintains two essential sea-level observation networks.

NOAA's National Data Buoy Center operates and maintains the U.S. network of Deep-Ocean Assessment and Reporting of Tsunami (DART) systems, which were developed by NOAA's Pacific Marine Environmental Laboratory for the early detection, measurement, and real-time reporting of tsunamis in the open ocean. The U.S. network is part of a larger international network.

Closer to shore, networks of coastal water-level stations are used to confirm tsunami arrival time and height. In the United States, most of these stations are operated and maintained by NOAA's Center for Operational Oceanographic Products and Services. Several others are operated by the tsunami warning centers.

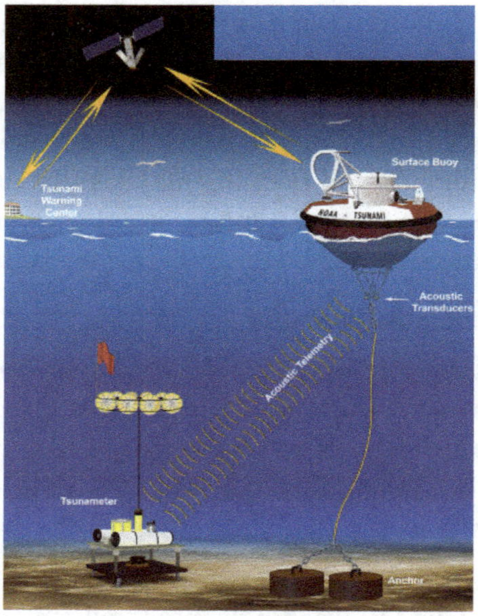

Fig. 5. This diagram shows how tsunami wave information in the deep ocean is transmitted from DART systems via satellite to NOAA's tsunami warning centers

Tsunami forecasting and warning

If an earthquake meets certain criteria, the warning centers issue U.S. messages (alerts) through multiple channels to emergency managers, other officials, news media, and the public. They issue threat messages to international partners in the Pacific and Caribbean to help them understand the threat to their coasts so they can determine the level of alerts to issue.

There are four levels of tsunami alerts in the United States: Information Statement, Watch, Advisory, and Warning. When the centers issue Tsunami Warnings, they are broadcast through local radio and television, wireless emergency alerts, NOAA Weather Radio and NOAA websites. They may also come through outdoor sirens, local officials, text message alerts, and telephone notifications.

If a tsunami is detected, the warning centers run tsunami forecast models developed by NOAA's Pacific Marine Environmental Laboratory and the warning centers. These models use real-time information from the observation systems and pre-established scenarios to simulate tsunami movement across the ocean and estimate coastal impacts, including wave height and arrival times, the location and extent of coastal flooding, and event duration.

The resulting forecasts, combined with historic tsunami data and additional seismic analysis, help the warning centers decide whether to cancel messages or adjust them. In the United States, these forecasts also provide local officials with actionable information that can guide decisions about population evacuation, including pedestrian and traffic routes and beach and road closures.

To support forecast and warning capabilities, NOAA's Center for Coasts, Oceans, and Geophysics develops high-resolution coastal digital elevation models, which depict Earth's solid surface. The center also serves as the long-term archive for national and international tsunami data, a natural hazards image database, and the global historical tsunami database. It is used to identify regions at risk, validate tsunami forecast models, help position DART systems and coastal water-level stations, and prepare for future events.

HOW ARE WAVES GENERATED, AND HOW DO THEY MOVE?

The ocean is never still. Whether observing from the beach or a boat, we expect to see waves on the horizon. Waves are created by energy passing through water, causing it to move in a circular motion. However, water does not actually travel in waves. Waves transmit energy, not water, across the ocean and if not obstructed by anything, they have the potential to travel across an entire ocean basin.

Waves are most commonly caused by wind. Wind-driven waves, or surface waves, are created by the friction between wind and surface water. As wind blows across the surface of the ocean or a lake, the continual disturbance creates a wave crest. These types of waves are found globally across the open ocean and along the coast.

More potentially hazardous waves can be caused by severe weather, like a hurricane. The strong winds and pressure from this type of severe storm causes storm surge, a series of long waves that are created far from shore in deeper water and intensify as they move closer to land. Other hazardous waves can be caused by underwater disturbances that displace large amounts of water quickly such as earthquakes, landslides, or volcanic eruptions. These very long waves are called tsunamis. Storm surge and tsunamis are not the types of waves you imagine crashing down on the shore. These waves roll upon the shore like a massive sea level rise and can reach far distances inland.

The gravitational pull of the sun and moon on the earth also causes waves. These waves are tides or, in other words, tidal waves. It is a common misconception that a tidal wave is also a tsunami. The cause of tsunamis are not related to tide information at all but can occur in any tidal state.

Looking out at the ocean, one often sees a seemingly infinite series of waves, transporting water from one place to the next. Though waves do cause the surface water to move, the idea that waves are travelling bodies of water is misleading.

Waves are actually energy passing through the water, causing it to move in a circular motion. When a wave encounters a surface object, the object appears to lurch forward and upward with the wave, but then falls down and back in an orbital rotation as the wave continues by, ending up in the same position as before the wave came by. If one imagines wave water itself following this same pattern, it is easier to understand ocean waves as simply the outward manifestation of kinetic energy propagating through seawater. In reality, the water in waves doesn't travel much at all. The only thing waves do transmit across the sea is energy.

The idea of waves being energy movement rather than water movement makes sense in the open ocean, but what about on the coast, where waves are clearly seen crashing dramatically onto shore? This phenomenon is a result of the wave's orbital motion being disturbed by the seafloor. As a wave passes through water, not only does the surface water follow an orbital motion, but a column of water below it (down to half of the wave's wavelength) completes the same movement. The approach of the bottom in shallow areas causes the lower portion of the wave to slow down and compress, forcing the wave's crest higher in the air. Eventually this imbalance in the wave reaches a breaking point, and the crest comes crashing down as wave energy is dissipated into the surf.

Where does a wave's energy come from? There are a few types of ocean waves and they are generally classified by the energy source that creates them. Most common are surface waves, caused by wind blowing along the air-water interface, creating a disturbance that steadily builds as wind continues to blow and the wave crest rises. Surface waves occur constantly all over the globe, and are the waves you see at the beach under normal conditions.

Adverse weather or natural events often produce larger and potentially hazardous waves. Severe storms moving inland often create a storm surge, a long wave caused by high winds and a continued low pressure area. Submarine earthquakes or landslides can displace a large amount of water very quickly, creating a series of very long waves called tsunamis. Storm surges and tsunamis do not create a typical crashing wave but rather a massive rise in sea level upon reaching shore, and they can be extremely destructive to coastal environments.

WAVE PROPERTIES AND CHARACTERISTICS

Material waves provide a mechanism for transferring energy over considerable distances, without the transport of the material medium itself.

In most material waves we typically encounter, the "shape" of the disturbance stays the same over short travel distances. In our example of a rock thrown into a pond, the ripples look similar as they expand away from the initial disturbance. Over greater distances, however, we notice changes. The ripples seem to die out as the radius of the circle they make increases This is because as the wave spreads out, its energy disperses. In some waves, the shape may remain constant over very long distances, for example low frequency sound waves at large depths of the ocean.

Pulse Wave

A pulse is a disturbance with a finite length. For example, if you shook the end of a rope once you would produce a pulse wave as shown in Fig 6.

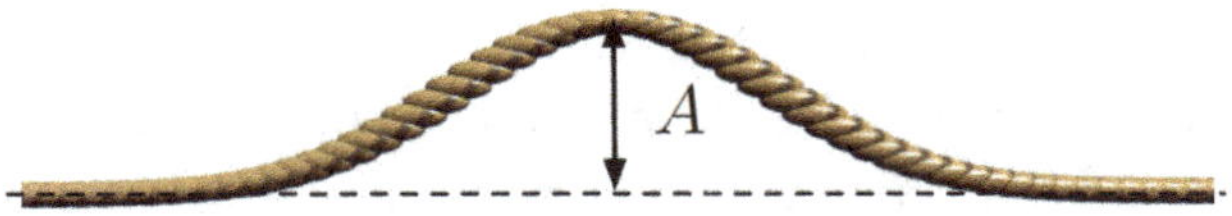

Fig. 6. Pulse Wave

The location of the rope before and after the pulse passes is the equilibrium position. The maximum magnitude of the displacement of the pulse from equilibrium is the amplitude, designated here by the letter AA. Note that as we've defined it, AA is always positive.

Another example of a pulse-type wave is the example of a rock thrown into a pond. In this example the rock produces many ripples, but they are a pulse wave because there are a finite amount of ripples with a well-defined beginning and end

A Repetitive (Periodic) Wave

Repetitive waves can have many different shapes. One of the simplest to deal with looks like a sine or cosine function. Such waves are called harmonic or sinusoidal waves, and are generated by oscillators moving in simple harmonic motion. For example, if you hold one end of a rope and jiggle it up and down in simple harmonic motion, you will generate harmonic waves. If you were to take a picture of the waves, it might look like this (Fig. 7).

We can represent this wave with a graph like the one below. Here x is the distance along the rope, and y is how far the rope is displaced from equilibrium. The curve is the shape of the rope at one particular instant t=0.

Fig. 7. Repetitive (Periodic) Wave

Like the wave pulse, a repeating wave has an equilibrium position (the location of the rope when no wave is present) and an amplitude A. In addition a repeating wave has two additional parameters. The first is the wavelength λ which tells us the distance (along the direction of wave motion) between identical pieces of the wave. The second is the period T; a wave completes one period if it enough time (T) has passed that the wave looks exactly the same. Pulse-type waves, which do not repeat, do not have periods or wavelengths. In other words, the wavelength λ tells us how the wave repeats in space, while the period T tells us how the wave repeats in time.

Notice that the graph above shows the wave for a fixed time, so it gives us no information about the period. For that we need a new graph; if we were to paint a red dot on the rope at some fixed x value, and then plot the position of only that dot against time, we would find that particular point moves in simple harmonic motion.

It takes the same amount of time for the dot to return to an initial position as it does for the whole wave to return to an initial configuration. That means TT is the period of both the simple harmonic motion of the red dot on the rope and of the wave itself. Just like in simple harmonic motion, the period is the reciprocal of the frequency:

$$T=1/f \qquad (1)$$

Typically, T is measured in seconds and f is measured in Hertz ($1\ Hz = 1\ s^{-1}$). T is the time between the arrival of adjacent crests of the wave, while f is the number of crests that pass by per second. This means the red dot completes an f number of up-and-down cycles every second, or the rope returns to an initial configuration an f number of times every second.

Dimensionality

The examples seen above are called one-dimensional waves because the displacement discussed in each only changes in one direction. We arbitrarily called that direction the y-axis.

If a wave spreads out on a surface then we will define it as a two-dimensional wave. For example a water wave spreads out on the surface of the water, so water waves are two-dimensional.

A wave that spreads outward in all directions is three-dimensional. Examples of three-dimensional waves are typical sound and light waves.

An important distinction between these waves is that, as we described them above, the amplitude A of the waves is only constant for one-dimensional waves. Two-dimensional and Three-dimensional waves have amplitudes that depend on the distance from the source of the disturbance. Recall how ripples in a pond become smaller as they spread out, or how the brightness and loudness of something increases as you get closer to it. This is a consequence of conservation of energy; as a wave propagates outward in multiple dimensions, the energy it carries must be spread over a region of increasing size.

HOW DOES THE WIND GENERATE WAVES?

The waves transport mass, and that wave-induced drift alters the dynamics of the upper ocean: It mixes the surface layers of water and modulates their temperatures, a crucial boundary condition between the air and sea in coupled models of Earth's weather and climate. More practically, statistical wave models describe how waves evolve and propagate across the ocean surface in various environmental conditions—such as low and high winds—and quantify how they influence the ocean's circulation and the transport of jetsam, flotsam, and pollution on its surface.

That wind produces waves in water is obvious to even the most casual observers. It has received attention for centuries from mathematicians and physicists. Even so, a complete description of the phenomenon eludes modern researchers. The difficulty comes from the fact that wave generation occurs at the interface of two fluids (here, air and water), with the flow in both generally turbulent. The waves vary sufficiently in space and time to be considered random, and the fluids interact over a broad range of scales—from millimeters to kilometers in space, and from seconds to hours in time. That range makes analytical and numerical progress extremely difficult.

The challenge in laboratory studies is to resolve the turbulent flow of air and water simultaneously over that large range on a curved and quickly changing surface. Field studies suffer from the difficulties of making measurements at sea, particularly when wind measurements are disturbed by the distortion produced by the presence of a research vessel. Despite those obstacles, much progress has been made experimentally and theoretically to understand the phenomenon. In this article, we outline the problem, review historical approaches to solve it, and discuss some of the open questions.

Wind-generation mechanisms in history

Many cultures have an intuitive understanding of the relationship between wind and waves. Micronesians and Polynesians, for instance, are famous for using swell to aid in their navigation. The modern treatment of the relationship began with two pivotal figures of 19th-century physics, Hermann von Helmholtz and William Thomson (later Lord Kelvin), who argued that wind generates waves through a shear-flow instability. The two scientists would often discuss the issue on trips out on Kelvin's boat.

The process, known as the Kelvin–Helmholtz instability, occurs whenever a fluid changes speed across a region of changing density. The pair calculated that for the mechanism to operate in realistic conditions, wind speeds of 6.5 m/s were needed to generate waves. But several laboratory experiments have since recorded wave generation at much lower wind speeds. Apparently, there is more to the story than what Kelvin and Helmholtz had proposed.

In 1925 Sir Harold Jeffreys argued that air flowing over water waves is, like air flowing past a sphere, deflected by surface geometry. The analogy led to an understanding of what's now known as airflow separation—a reversal in the direction of airflow over the leeward side of a wave crest (Fig. 8).

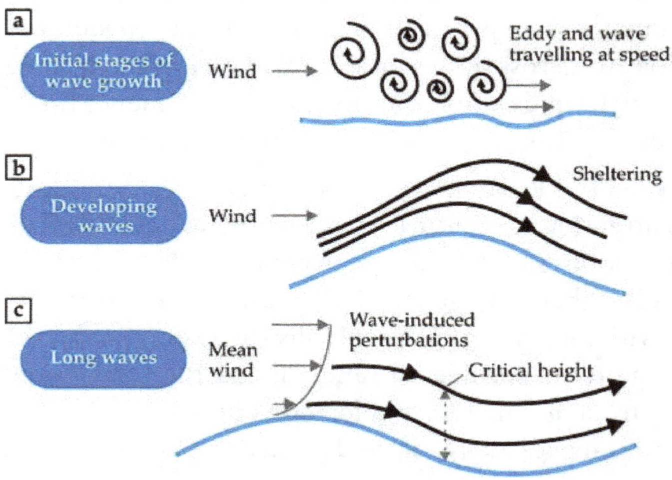

Fig. 8. Wave-generation mechanisms.

(a) Turbulent eddies in the air disturb an initially calm ocean and create ripples with wavelengths on the scale of centimeters. (b) Those ripples grow to meter-scale wavelengths, and the wind becomes "sheltered" on the downstream (leeward) side of the wave crest. The pressure difference between the windward (left) and leeward (right) sides of the crest transfers energy from the wind to the wave, causing it to grow. (c) The wind's speed is highest well above the water and decreases until it reaches zero at the ocean surface. At a critical height where the wind's speed equals the phase speed of the wave, the wind's shear resonates with the wave and transfers further energy to it.

The geometric phenomenon, which Jeffreys called sheltering, arises from a pressure difference between the windward and leeward sides of the wave. The wind pressure and slope of the water wave are both oscillatory, and when the two are phase-shifted with respect to each other, work is done on the wave, which causes it to grow. The theory has an unconstrained scaling parameter, known as the sheltering coefficient, that estimates the work done on the waves by the wind.

Preliminary laboratory experiments of wind over solid objects showed that the sheltering coefficient depends crucially on the specific geometry of the object, which the Jeffreys theory does not account for.

Current research challenges

Recent advances to understanding how wind generates waves have been driven by technological developments in computational and observational capabilities, and in improved theoretical formulations of the problem.

Measurements made in the laboratory and in the field have shown the existence of a critical height—a necessary feature of Miles's theory—and its importance in controlling the flux of momentum to long waves. Natural reference frames, which measure distances normal and tangential to the surface, not from some fixed point, have greatly clarified the observational data. The development of large-eddy simulations of the atmospheric boundary layer have also shed light on various regimes, particularly for low winds and long waves. In those regimes, significant momentum can even be transferred from the waves back to the atmosphere.

The dynamics of short waves are more complex than those of long waves because of the effects of wave breaking. The steep slopes that occur when waves break induce 3D airflow separation downwind of the wave crest (see Fig. 9).

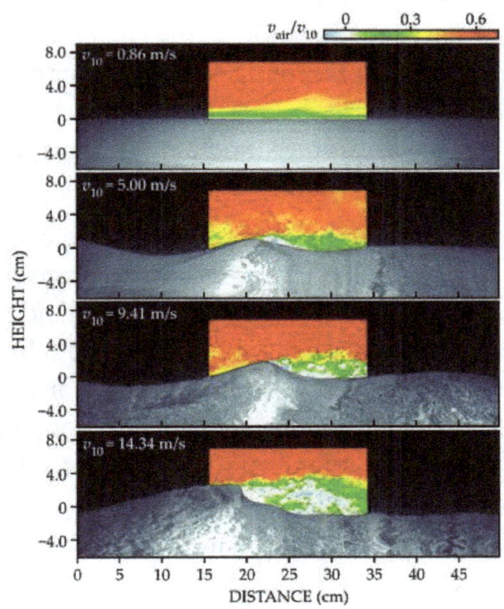

Fig. 9. Airflow speed, in color, as measured above water waves.

Wave-breaking events become even more important as the wind strength increases to hurricane levels. Experiments in laboratory facilities capable of producing such extreme conditions indicate that the reattachment of separated air streamlines occurs much less frequently at higher wind speeds than at lower ones and isolates the waves from the bulk airflow. That modification of properties in the near-surface flow has proven to be important in hurricane-strength conditions and is an active area of research.

Some experiments and computations that resolve the fully coupled air–water turbulent system have already been achieved and are starting to become more routine. Experimentally, the challenge is to perform measurements of the turbulence in air and water simultaneously. Computationally, the challenge is to solve the two-phase Navier–Stokes equations over the wide range of scales that are involved.

Figure 9 shows snapshots of the airflow in recently performed laboratory experiments just above water waves of various slopes and in increasing wind intensities.

SURF ZONE

The surf zone is one of the most dynamic regions of the marine environment. Not only is it highly energetic, it also supports a wide diversity of life forms. Surf zones in both sandy and rocky shores produce dynamic surroundings that shape the entire coastal region.

The surf zone is a complex place. Breaking waves create currents that run offshore (rip currents) and other currents that flow along the beach (alongshore flows). The direction that incoming waves travel changes as the water depth changes. The waves are reflected and refracted because the speed at which the waves move depends on water depth. (Ocean surface waves have much in common with sound waves, which are reflected).

In particular, submarine canyons located near shore can have strong effects on the incoming waves and the surf zone. Submarine canyons are narrow, steep-sided valleys on the sea floor that resemble river canyons on land. The Nearshore Canyon Experiment was conducted off San Diego, where two deep submarine canyons come in close to shore. Scientists measured how submarine canyons affect wave propagation and nearshore currents. A large number of Acoustic Doppler Velocimeters (ADVs) and Acoustic Doppler Current Profilers (ADCPs), as well as many other instruments, were deployed.

Waves are initiated at sea where winds sweep across the surface of the water . Some of the energy of the wind is transferred to the surface and small sinusoidal waves appear on top of the water. The wave often increases in energy with added winds. The height of oceanic waves is dependant on the energy of the wind which may be quite high during storm events.

The waves continue until they reach the shore. Once there, they strike the shore and lose their energy. What happens during this process is of great importance to geologists and those who live near the beach.

Ocean waves are very interesting examples of wave **physics** found in nature. A look at the wave in cross-section appears as a series of linked "S" shapes. The high point of the individual curve is called the crest. The low point is the trough. The size of the individual waves (the amplitude) and the frequency (the rate at which a series of waves pass a particular point during a specific amount of time) are determined by the amount of energy a group of waves contains.

The long waves strike the shore at an angle. Waves will rarely, if at all, touch the shore at a parallel line to the shore. This is a result of the changing topography of the land. When the wave strikes the shore it is refracted and a great deal of its energy is directed at an oblique angle to the original wave. This increases the energy of the wave effect and, for anyone who has experienced it, produces a significant push-pull type of action. Being caught in this area , named the swash, can be dangerous under certain conditions such as rip tides formation.

Geologists are concerned about the region where waves break on the shore. Small circular currents occur in the water that produce the characteristic swell shape. As these circular currents reach the shore and strike the bottom, they are compressed. This compression eventually breaks down the wave structure and the water spills over on itself at the crest. This is the point at which waves are commonly called "breakers." The bottom sediments act as a further drag on the wave and deplete it of much of its energy. Any sediments that were initially captured by the breaking wave are dumped and moved onto the beach. At the same time, the refracted wave captures sediment and pulls it back down the shoreline. This repeated motion causes **sand** grains to migrate down the beach as waves strike again and again. Where the shore is rocky, the force of the waves wears away the rocks and transports smaller clasts (sedimentary particles) out to sea and the coast. This type of **erosion** is a major physical process that sculpts cliffs around the world. The entire area of breaking waves, including the slope-face of the land, is called the surf zone.

Even though they bend and become nearly parallel to shore, most waves still reach the shore at a small angle, and as each one arrives, it pushes water along the shore, creating what is known as a longshore current within the surf zone (the areas where waves are breaking) (Fig. 10).

Another important effect of waves reaching the shore at an angle is that when they wash up onto the beach, they do so at an angle, but when that same wave water washes back down the beach, it moves straight down the slope of the beach (Fig. 11). The upward-moving water, known as the **swash**, pushes sediment particles along the beach, while the downward-moving water, the **backwash**, brings them straight back. With every wave that washes up and then down the beach, particles of sediment are moved along the beach in a zigzag pattern.

The combined effects of sediment transport within the surf zone by the longshore current and sediment movement along the beach by swash and backwash is known as **longshore drift**. Longshore drift moves a tremendous amount of sediment along coasts (both oceans and large lakes) around the world.

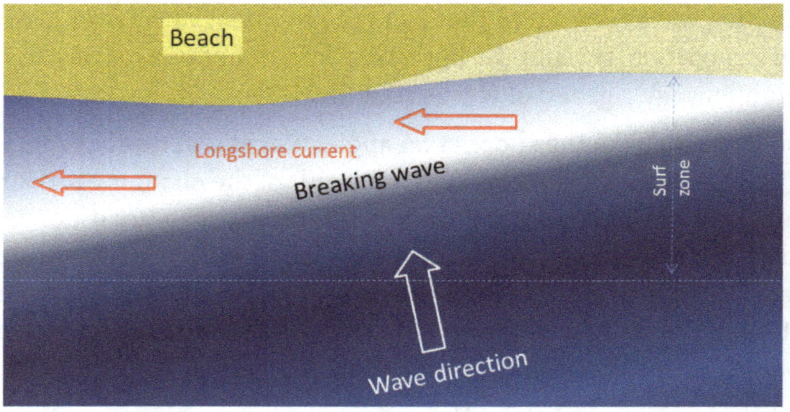

Fig. 10 The generation of a longshore current by waves approaching the shore at an angle.

A **rip current** is another type of current that develops in the nearshore area, and has the effect of returning water that has been pushed up to the shore by incoming waves. As shown in Fig. 12, rip currents flow straight out from the shore and are fed by the longshore currents. They die out quickly just outside the surf zone, but can be dangerous to swimmers who get caught in them. If part of a beach does not have a strong unidirectional longshore current, the rip currents may be fed by longshore currents going in both directions.

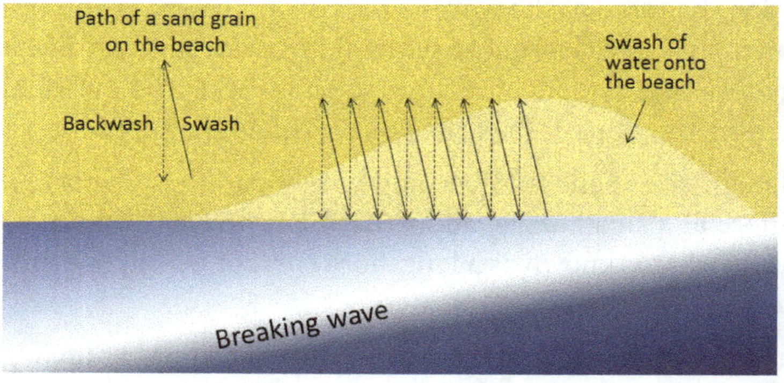

Fig. 11 The movement of particles on a beach as a result of swash and backwash

Tides are related to very long-wavelength but low-amplitude waves on the ocean surface (and to a much lesser extent on very large lakes) that are caused by variations in the gravitational effects of the Sun and Moon. Tide amplitudes in shoreline areas vary quite dramatically from place to place. On the west coast of Canada, the tidal range is relatively high, in some areas as much as 6 m, while on most of the east coast the range is lower, typically around 2 m. A major exception is the Bay of Fundy between Nova Scotia and New Brunswick, where the daily range can be as great as 16 m. Anomalous tides like that are related to the shape and size of bays and inlets, which can significantly enhance the amplitude of the tidal surge. The Bay of Fundy has a natural oscillation cycle of 12.5 hours, and that matches the frequency of the rise and fall of the tides in the adjacent Atlantic Ocean. Ungava Bay, on Quebec's north coast, has a similarly high tidal range.

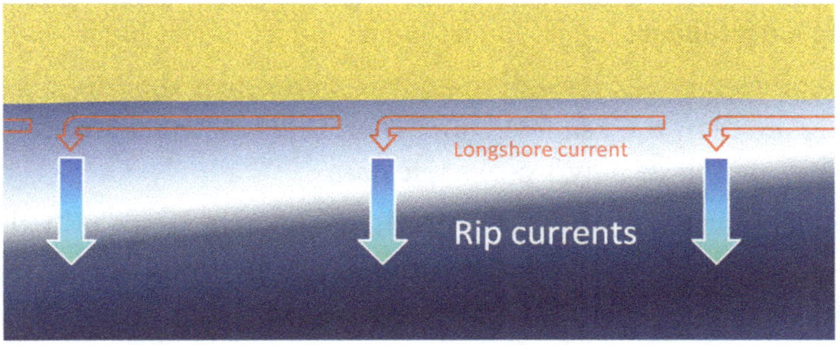

Fig. 12 The formation of rip currents on a beach with strong surf

HOW DO WE HARNESS WAVE ENERGY?

There are many ways of classifying the various types of technology used to convert wave energy to electricity. For simplicity, they can be divided into three main types: oscillating water columns, oscillating bodies and overtopping convertors.

Oscillating Water Columns

Oscillating water columns use wave surges to drive a stream of trapped air through an air turbine. These devices are essentially large hollow columns half submerged in the ocean, with an underwater vent open to the sea. The movement of waves pushes water into the chamber, forcing air through the column. As waves recede, air is sucked back through the column in the other direction.

The movement of air rotates a turbine in the column, which rapidly turns a generator to produce electricity. The turbines in oscillating water columns are specially designed to turn in the same direction irrespective of the direction of the air movement, which takes advantage of the back-and-forth movement of air.

Oscillating Bodies

Oscillating bodies use floating buoys or platforms that rise and fall with the swell. There is no one standard for this type of technology, and they appear in many forms.

The most common oscillating bodies are called point absorber buoys. They are fixed to the sea floor via a hydraulic pump. The buoy moves up and down along ocean swell crests and troughs, activating the hydraulic pump which pushes water or air through a turbine, which in turn rotates a generator to produce electricity.

Another commonly used wave power generator is a surface attenuator. This is a series of long floating cylindrical sections, connected by hinges, that are arranged perpendicular to oncoming waves. As the wave passes along the attenuator, the cylindrical sections move up and down in relation to each other. This causes hydraulic pumps at the hinges to 'pinch', pumping high pressure fluid to rotate a generator and produce electricity.

Overtopping Converters

Overtopping converters allow swell to deliver water over the top of a structure that captures it higher than the sea level, and then releases the water to drive a hydro-electric turbine, similar to a hydro-electric dam.

Ocean swells that would normally crash against the shore or a cliff are instead directed up a ramp to a water tank approximately 3-5 metres above sea level. The water is then released, through a turbine, back to sea level.

Modern wave power did not emerge until the 1970s, when the world experienced a significant oil crisis. This drove many nations to explore alternatives to oil to reduce dependency on the diminishing and fluctuating resource. Despite high interest and investment in the 1970s, funding for wave power dropped as oil prices returned to normal levels.

However, research is ongoing and several countries with coastlines, including Australia, Canada, the USA, the UK, Ireland and some European countries are leading the development of wave energy technologies.

Technology

There are multiple different technologies used for Wave energy. There are five main types of technology used including; Absorbers, Attenuators, Oscillation water columns, overtopping and Inverted- Pendulum device.

Abosorbers extract energy from the rise and fall of the waves with a buoy. Once the energy is extracted it is then converted to electrical energy with a linear or rotary generator.

Current and Possible Wave Farms

Since wave energy is still relatively new, there are fewer areas that currently have or are installing wave farms. Some main areas for wave farms are Portugal, the United Kingdom, Australia, and the United States.

Portugal has the very first wave farm, the Aguçadoura Wave Farm. It's about 3 miles offshore, north of Porto, and was designed to use 3 Pelamis wave energy converters. It has an installed capacity of 2.25 MW, and was officially opened in September, 2008. Due to the financial collapse from an economic crisis, however, it was shut down.

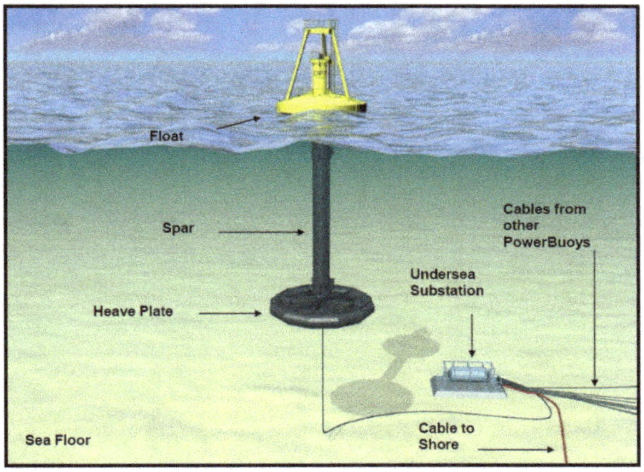

Fig. 13. Attenuators capture energy by being placed perpendicular to the length of the wave, this causes the attenuator to contentiously flex where segments are connected. This connection is then connected to hydraulic pumps which convert the energy

United Kingdom. In May, 2010, a wave farm (currently 66 machines) was launched in Scotland. It was announced in 2007 and cost over 4 million pounds, equivalent to $5.9 million dollars.

Off the north coast of Cornwall, England, the Wave hub facility acts as a giant extension cable. It allows 20 MW to be connected to it, but could go up to 40 MW. It also could end up powering up to 7500 households and save around 300,000 tons of CO2 in the next 25 years.

Australia have wave farms that are currently under development off the West coast and near Portland, Victoria. Both of these projects are targeted to cost upward of $65 million. There is also a *greenWave* device off of Southern Australia.

United States The United States was to have a wave farm off of the West coast by Oregon. The Reedsport farm was supposed to be installed in 2013, but due to legal and technical problems, the project came to a halt.

CHAPTER

9 TIDES

Tides can be defined as the alternate rise and fall of the ocean water. It is caused by the combined effects of :

1. The gravitational force exerted on Earth by the Sun
2. The gravitational force exerted on Earth by the Moon
3. Rotation of the Earth

What are the types of tides & what causes these tides?

TYPES OF TIDES

- **Tides Based on Frequency**
  - Semi-Diurnal Tides
  - Diurnal Tides
  - Mixed
- **Tides Based on the Position of Earth, Sun, and the Moon**
  - Spring Tides
  - Neap Tides

Semi-Diurnal Tides

A semi-diurnal tidal cycle is the one with two nearly equal high tides and two low tides each day. The interval between the high and the low tides is of around 12 hours and 25 minutes. Semi-Diurnal Tides are most widespread in the Indian Ocean. The other prevalent coasts where semi-diurnal tides are experiences are:

- Eastern African Coast
- Bay of Bengal

Diurnal Tides

It means four tides in a day. Two tides by the sun and two by the moon. Spring Tide It is an exceptionally high tide generated by the complementary factor played by the Sun with respect to the moon. It should be noted that when Sun, Moon, and Earth are in the same line, the position is known as the Syzygy. This syzygy can be of two types :

1. Conjunction: when the moon and sun are on the same side
2. Opposition: When the moon and sun are on the opposite side In both of these conditions, the magnitude of the tide will be equally high.

Mixed Tides

A tidal cycle with two unequal high and low tides lead to the formation of the mixed tidal cycle, or simply called mixed tide. This tidal cycle has both semi-diurnal and diurnal oscillations. It is widely observed in the Gulf of Mexico and the Caribbean Sea. Southeastern Brazilian coast also witnesses mixed tides.

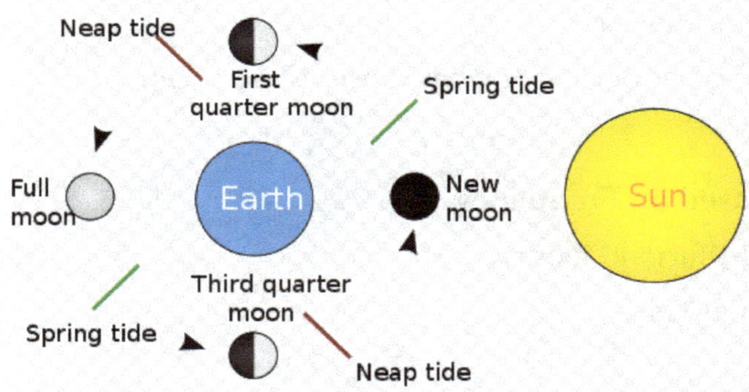

Fig. 1. Mixed Tides

Spring Tides

Spring tides are formed when the sun and the moon are in line with each other and pull the ocean surface in the same direction. This leads to higher high tides and lowers low tides and such tide is called a spring tide. In a lunar month, it occurs twice. It is also known by the name of 'King Tide.'

Note: The aspirants should know that the spring season has nothing to do with spring tides. The word 'Spring' in spring tides means 'springing forth.' These occur in full or new moon days. In both new moon or full moon days, the sun's gravitational pull is added to the moon's gravitational pull on Earth, causing the oceans to bulge a bit more than usual. This results in 'higher' high tides and 'lower' low tides.

Neap Tides

It occurs seven days after the spring tide. The prominent point is that the sun and the moon are at the right angle to each other. This tide occurs during the first and the last quarter of the moon. The gravitational pull of the moon and the resulting oceanic bulge is cancelled out by the gravitational pull of the sun and its resulting oceanic bulge. Also, in contrast to spring tides, the high tides are 'lower' and the low tides are comparatively 'higher' in neap tides.

HOW DO TIDES WORK?

The mechanism of tides could be understood by understanding the gravitational force of the Sun and the Moon. These bodies experience the gravitational pull over each other depending upon their mass and the distance between them. Since the Sun is far away from the Earth as compared to the Moon. Hence, the Sun's gravitational pull is lesser over the Earth than the moon. Thus, the moon determines the magnitude of the tide. It is supposed that only the water bodies are pulled by the gravitational pull, however, it is not the fact. It is both the land and water bodies that get pulled by the gravitation. Since the relative pull of the land is less in comparison to that of water, the effect of gravitation on the water bodies is more. It should be noted that the magnitude of ant tide is determined by the relative position of the Moon, the Sun, and the Earth.

Impact of Tides

1. Tides raise the level of seawater and hence exposes a large part of the ocean for erosion
2. It is helpful for the tidal ports that have shallow water which is a constraint for the big ships to enter. Tidal currents are a very potential source of tidal energy which is harnessed by many developed countries on a very large scale and to some extent in India as well. It can be devastating in cases where the tide gets too huge and results in the flooding of the nearby coastal regions. Tides are very helpful for ecosystems such as the mangrove forests and coral reefs to grow and sustain.

CHARACTERISTICS OF TIDES

Tide changes as per the 4 stages given below.

1. **Flood Tide** – Over a period of several hours there will be a rise in sea level.
2. **High Tide** – This is a stage where the water reaches its maximum level.
3. **Ebb Tide** – This is a stage where sea level keeps receding over several hours.
4. **Low Tide** – The Level of Seawater stops receding.

WHAT IS A TIDAL BULGE?

The side of the earth that is nearest to the moon witnesses one tidal bulge. The other side of the earth that is farthest to the moon witnesses the second tidal bulge. These tidal bulges are high tides. The water in the ocean is pulled towards the moon under the impact of the gravitational pull. That creates a tidal bulge.

WHAT CAUSES OCEAN TIDES?

High and low tides are caused by the moon. The moon's gravitational pull generates something called the tidal force. The tidal force causes Earth—and its water—to bulge out on the side closest to the moon and the side farthest from the moon. These bulges of water are high tides.

High tides and low tides are caused by the moon. The moon's gravitational pull generates something called the tidal force. The tidal force causes Earth—and its water—to bulge out on the side closest to the moon and the side farthest from the moon. These bulges of water are high tides.

As the Earth rotates, your region of Earth passes through both of these bulges each day. When you're in one of the bulges, you experience a high tide. When you're not in one of the bulges, you experience a low tide. This cycle of two high tides and two low tides occurs most days on most of the coastlines of the world.

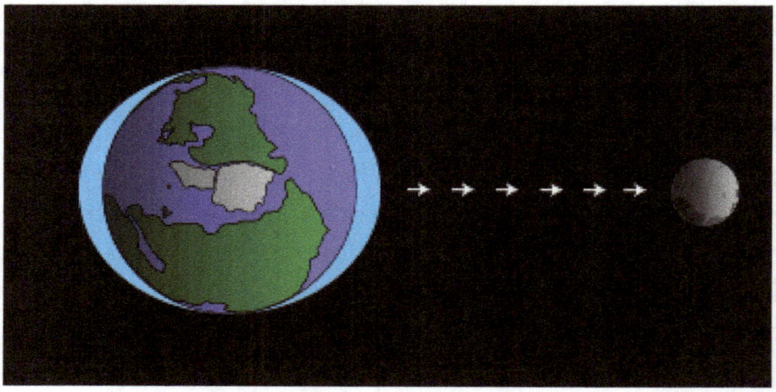

Fig. 2. The tidal force causes water to bulge toward the moon and on the side opposite the moon. These bulges represent high tides.

Tides are really all about gravity, and when we're talking about the daily tides, it's the moon's gravity that's causing them.

As Earth rotates, the moon's gravity pulls on different parts of our planet. Even though the moon only has about 1/100th the mass of Earth, since it's so close to us, it has enough gravity to move things around. The moon's gravity even pulls on the land, but not enough for anyone to tell (unless they use special, really precise instruments).

When the moon's gravity pulls on the water in the oceans, however, someone's bound to notice. Water has a much easier time moving around, and the water wants to bulge in the direction of the moon. This is called the tidal force.

Because of the tidal force, the water on the side of the moon always wants to bulge out toward the moon. This bulge is what we call a high tide. As your part of the Earth rotates into this bulge of water, you might experience a high tide.

The ocean also bulges out on the side of Earth opposite the moon.

HOW DO TIDES VARY DURING A MONTHLY TIDAL CYCLE

The monthly tidal cycle is days because that's how long it takes the Moon to complete an orbit around Earth. During its orbit around Earth, the Moon's changing position influences tidal conditions on Earth.

The Monthly Tidal Cycle During the monthly tidal cycle, the phase of the Moon changes dramatically. When the Moon is between Earth and the Sun, it cannot be seen at night, and it is called the new moon.

When the Moon is on the side of Earth opposite the Sun, its entire disk is brightly visible, and it is called a full moon. A quarter moon—a moon that is half lit and half dark as viewed from Earth—occurs when the Moon is at right angles to the Sun relative to Earth.

Both the moon and sun affect the tides, but since the moon is much closer to the earth (384,400 km instead of 149,600,000 km), it has more than twice the effect of the sun, even though it is much smaller. So, to understand tides it's best to start with the moon and the lunar tide.

The earth isn't fixed rigidly in space, and as the moon orbits, it attracts the earth round in a monthly orbit of its own. So the earth has a small orbit caused by the moon in addition to its annual one round the sun. It is the gravitational pull of the moon on the earth which keeps the earth in its monthly orbit.

This is similar to when you whirl a weight round on a piece of string. It is only by constantly pulling on the string that you prevent the object from flying off. If you stop pulling completely, by letting go on the string, the object does fly off. Gravity acts in the same way as you pulling on the string, and prevents the earth from flying off. However, unlike the string, the attractive pull of gravity gets weaker as the distance between the objects gets larger (and it becomes stronger, of course, the closer they get). The earth is large, so the pull of gravity on the side of the earth nearer the moon is stronger than the pull on the side of the earth farther from the moon.

The pull of the moon's gravity is just enough to keep the earth in its monthly orbit, but it is a bit stronger on the surface of the earth facing the moon (near side) and weaker on the far side. This means that on the near side, the moon tends to pull anything that's free to move towards it. In the same way, on the far side, there isn't quite enough gravitational pull, so that anything that's free to move tends to fly off, away from the moon. So on the near side they are pulled into a bulge towards the moon, and on the far side, they pile up into a bulge away from the moon.

As the earth spins, different parts of the world move under the two bulges of high water and experience high tides, giving the familiar two tides a day around Britain. In some parts of the world, local effects can mean only one tide a day, or even none.

Between the two bulges are two troughs of low water, producing two low tides a day. The sun also creates a very similar though smaller effect (the solar tide) and it is the interaction of the lunar and solar tides that causes spring and neap tides.

Springs and neaps

Tides change in height – low water level and high water level vary throughout the month. The tides build up to a maximum and fall to a minimum twice a month. The tides with the biggest difference between high and low water are called springs and those with the smallest are called neaps.

Spring tides happen just after every full and new moon, when the sun, moon and earth are in line. That's when lunar and solar tides line up and reinforce each other, making a bigger total tide. Neap tides occur when the moon is in the first or third quarter - when the sun, earth and moon form a right angle. The lunar high tide coincides with the solar low tide and they partly cancel out, giving a small total tide.

The regular motion of the sun, moon and earth cause spring tides to occur roughly 36 to 48 hours after the full or new moon, and for any given location, always at roughly the same time of day. For example at Liverpool, the spring tides are generally around midnight and midday.

Not all spring tides are the same size. Springs nearest the equinoxes (21 March and 21 September – when day and night are of equal length all over the world) are slightly bigger. The explanation for this is very complex and won't be given here.

Worldly effects

Back on earth, tides are affected by more mundane things, such as weather and the shape of the coastline. In the upper reaches of estuaries the river bed and shape of the estuary can distort the tidal pattern, producing long flood times and short ebb times. Though the time from high tide to high tide is the same as usual – roughly 12 hours 25 minutes – low tide is not half way between them. At some places, the flood tide rises quickly after a period of low water lasting maybe four or five hours. This kind of geographical effect is very important, and because of it you should use local tide tables (available from local newsagents and tackle shops) when possible.

Cromer, Lowestoft and Aldeburgh are fairly close to each other on the East Anglian coast. However, high tide is hours apart at these places. This is precisely the opposite to Southend and Herne Bay, which face each other across the mouth of the Thames. Their tides differ by only a few minutes. Without detailed geographical information, it is impossible to predict these differences, so keep to local tide tables.

Geography also affects the tidal range. Looking at tide tables for all of Britain, it's clear that the height of the tide varies around the country. For example the spring tidal range at Avonmouth is 12.2 m while at Lowestoft it's only 1.9 m.

Weather which can have a profound effect on the tide, is impossible to predict when calculating tide tables. Strong winds and abnormal atmospheric pressure are two of the main causes of altered tides. For example, a strong wind blowing on to land has the effect of piling up the water, giving a higher than predicted tide.

Moon Plays the Biggest Role

While both the Moon and the Sun influence the ocean tides, the Moon plays the biggest role. Although the Sun's gravitational pull on the Earth is 178 times stronger than the Moon's, the tidal bulges it causes are much smaller.

This is because, contrary to common belief, tides are not caused by the gravitational forces of the Moon or the Sun *lifting up* the oceans—their gravitational pull is much too weak for that. Rather, tides are created because the strength and direction of the gravitational pull *varies* depending on where on Earth you are. This variation creates the differential forces or tidal forces that in turn cause tides.

The tidal forces of the Moon are much stronger than the Sun's because it is so much closer to our planet, causing a much greater *variation* in the gravitational force from one location to another. The Sun's gravitational force, on the other hand, varies much less because the Sun is so far away.

The Oceans Bulge

The overall effect of these tidal forces is to "squeeze" the oceans, and produce two tidal bulges on opposite sides of the Earth - one facing the Moon and a slightly smaller one facing away from the Moon. Due to Earth's rotation, the two bulges act like two expansive "waves" continuously undulating around our planet.

Topography Causes Variation

Mid-ocean, each tidal "wave" is just under a meter high, compared to the water level of the two troughs between them. However, the variation between high and low tide is very different from place to place. It can range from almost no difference to over 16 meters (over 50 feet).

This is because the water in the oceans is constrained by the shape and distance between the continents as well as varying ocean depths. As a result, the tides behave more like water sloshing around in an oddly shaped bathtub than in a smooth and even basin. In some places, the water flows freely and quickly, while in other areas, where the water has to pass through narrow channels, it moves more slowly.

High and Low Nearly Twice a Day

Tides are one of the most reliable phenomena in the world, and we know that they move in and out around twice a day, but not exactly. So, why is that? A day on Earth is the time it takes our planet to spin once around its own axis in relation to the Sun. This is known as a solar day, and it lasts around 24 hours.

However, the time it takes Earth to reach the same position in relation to the Moon is, on average, 24 hours and 50 minutes, known as a lunar day. The reason the lunar day is longer than a solar day is that the Moon revolves around Earth in the same direction as Earth rotates around its axis, so it takes Earth, on average, an additional 50 minutes to "catch up" to the Moon. Because the tidal force of the Moon is more than twice as strong as the Sun's, the tides follow the lunar day, not the solar day. It takes half a lunar day, on average 12 hours and 25 minutes, from one high tide to the next, so we have high and low tides nearly twice a day.

According to the National Ocean Service, there are some exceptions to the main rule of two tides every lunar day. Along the coastline of the Gulf of Mexico, there is only one tide per day due to the local shoreline topography, among other things. This tidal cycle is called a *diurnal* cycle, as opposed to the normal *semidiurnal* cycle, where *diurnal* means daily and *semi* means *half.*

Continents Affect Tidal Lag

While in theory, the tidal bulges follow the Moon's position on its orbit around the Earth, the depth and shape of the ocean and the distance between continents are also important in determining when the tide rolls in and out. The time that passes between the passage of the Moon and the onset of the high tide is called the tidal lag.

In the Southern Ocean, where tidal bulges can move relatively freely, the tidal lag may be around two hours. On the other hand, the tidal lag in the North Sea—a part of the Atlantic Ocean bounded by continental Europe and the British Isles—can be about two days.

Gradual Ebb and Flow

The change from low to high tide is known as *flood tide*, while the change from high to low tide is called *ebb tide*. The technical term for the difference in water level between high tide and low tide is *tidal range*.

The flow and ebb are gradual, so it is not accurate to say that a high or low tide lasts around 6 hours and 12 minutes, i.e. a quarter of a lunar day. The speed of the water flow varies during this period, and it also varies from place to place.

The Rule of 12ths

People who have to consider the tides in their daily life, like sailors, fishers, and surfers, often use what is called the rule of 12ths to calculate the expected water level.

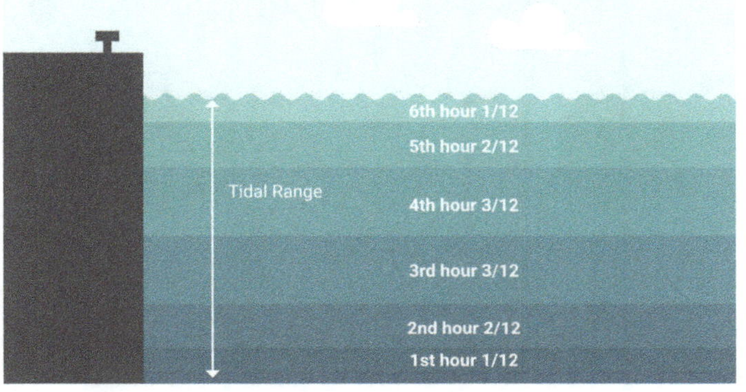

Fig. 3. The Rule of 12ths: The water level changes gradually

This rule states that in the 1st hour after low tide the water level will rise by 1/12 of the predicted tidal range in any given area. In the 2nd hour, it will rise 2/12, and in the 3rd hour, it will rise 3/12. In the 4th hour, it will also rise 3/12, in the 5th, it will rise 2/12, and in the 6th hour, it will rise 1/12. The sequence to remember is 1-2-3-3-2-1.

So, let's say the predicted tidal range is 12 feet. In the 1st hour, the tide would rise 1 foot. In the 2nd hour, it would rise 2 feet. In the 3rd and 4th hours, it would rise 3 feet. In the 5th hour, the tide would rise 2 feet, and in the 6th hour, 1 foot.

DYNAMIC THEORY OF TIDES

The Equilibrium Theory of tides predicts that each day there will be two high and two low tides, each one occurring at the same time day after day, with each pair producing tides of similar heights. While this view provides a basic explanation for the primary forces that generate the tides, it does not take into account such variables as the effects of the continents, the depth of the water, and many other factors.

The Dynamic Theory of tides takes these other factors into account, and shows that the tides are much more complicated and variable from place to place than the Equilibrium Theory would suggest. For example, some areas receive only one high and one low tide per day. Furthermore, the tidal range varies greatly across the globe; in the Mediterranean Sea, there can be a difference of only 10 cm between high and low tides, while the Bay of Fundy in Canada experiences a tidal range of up to 17m (56 ft) every day.

Examination of any tide chart will show that the tides don't occur at same time each day; in fact, each tidal peak occurs about 50 minutes later than it did in the previous day. This is due to the orbit of the moon around the Earth. Imagine a high tide that occurs at a particular location (X) at 1:00 pm (Fig. 4).

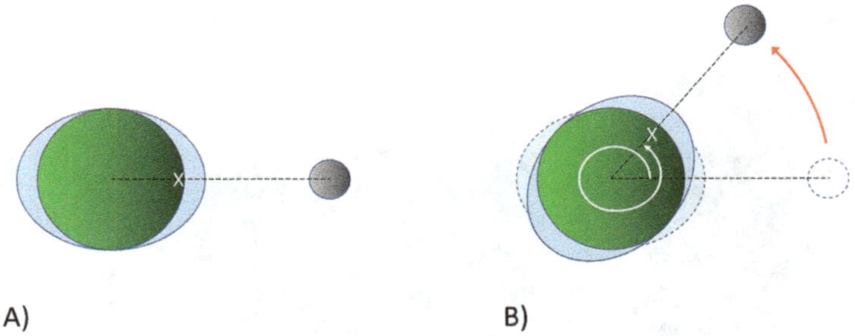

Fig. 4 In A) a high tide is occurring at point X on Earth's surface. 24 hours later, X has made a complete rotation and is back in its original position. However, the moon has moved during that time (B), so X must travel an additional distance (white arrow) to once again become aligned with the moon and experience a high tide. For this reason, corresponding tides occur approximately 60 minutes later each day

The high tide occurs as location X moves through the bulge of water facing the moon. It will take the Earth 24 hours to complete one revolution, to bring location X back to site of the water bulge that caused that high tide. However, during those 24 hours, the moon has also moved as it orbits the Earth, so the high tide bulge has moved beyond its original location. The Earth thus has to rotate an additional distance for location X to reach the bulge and experience that same high tide. Because it takes the moon about 28 days to orbit the Earth, the moon gets "ahead" of the Earth's rotation by about 50 minutes per day. Therefore, it takes location X 24 hours and 50 minutes to rotate through the same tidal bulge, and as a result, the tidal peaks occur about 50 minutes later each day. In our example, an afternoon high tide at 1:00 pm on one day would be followed by a high tide at about 1:50 pm the following day. This 24 hour and 50 minute cycle is referred to as a tidal day.

The motion of the moon impacts the tidal cycles in other ways. As the moon orbits the Earth, its orbital plane is at an angle relative to the rotational plane of Earth. This angle, or **declination**, means than the moon fluctuates between an angle of 28.5° north of the equator, to 28.5° south of the equator roughly every two weeks (the cycle from maximum to minimum and back takes about 27 days). A point on the Earth at the latitude indicated by the red line would experience two high tides as it rotated through 24 hours, at points A and B. But the two high tides would not be of equal heights; the high tide at A would be higher than the high tide at B. This helps create a mixed semi-diurnal tide; two high tides of different heights per day.

Finally, the continents and the bottom topography of the oceans have an impact on the tides that are experienced in an area. Because the tides are essentially waves with extremely long wavelengths extending halfway across the Earth, they behave as shallow water waves, and they are influenced and refracted by the bottom contours, leading to regional tidal variations. When the tidal crests encounter land, they are are reflected, and the wave moves back out to sea, theoretically until it encounters another continent on the opposite side of the ocean basin.

The crest is once again reflected, and the water oscillates back and forth as a standing wave across the ocean basin. However, because of the scale over which these tidal waves move, we must take into account the influence of the Coriolis Effect. As the tidal crest is reflected back across the ocean basin, its path is deflected by the Coriolis force; to the right in the Northern Hemisphere, and to the left in the Southern Hemisphere.

Using the Northern Hemisphere as an example, imagine a tidal crest that has reached land on the western side of an ocean basin. It would have a tendency to be reflected and move across the basin towards the east. But the Coriolis force deflects the movement to the right, causing the crest to instead head south. When the crest hits land in the south, it would now tend to reflect towards the north, but once again the Coriolis deflection to the right kicks in, and the wave instead moves to the east. From the east the reflected wave is deflected to the north, and so on. The result of all of this is that instead of a simple standing wave moving back and forth across the ocean, the tidal crest follows a circular pattern around the ocean basin, counterclockwise in the Northern Hemisphere and clockwise in the Southern Hemisphere.

This is analogous to shaking a pan full of water in a circular manner, and watching the water follow a similar circular path as it sloshes around inside. This large scale circular rotation pattern of tides is called amphidromic circulation. The rotation occurs around a central amphidromic point or node, that shows little tidal variation, while the largest tidal ranges occur on the edges of the circulation pattern.

In Fig. 5 the amphidromic points are indicated by the dark blue areas where the white lines converge, like spokes from a bicycle wheel, and the dark red and brown areas show the regions of maximum tidal heights. The tidal maxima will rotate around the amphidromic points, taking about 12 hours for a complete rotation, leading to two high and two low tides per day in many places. If a tidal maximum is occurring along one of the white lines in Fig. 5 at a certain time in the Northern Hemisphere, one hour later that high tide will have moved to the white line to the left (counterclockwise), and so on until it completes a rotation. In the Southern Hemisphere, the tide will move to the line to the right for clockwise rotation.

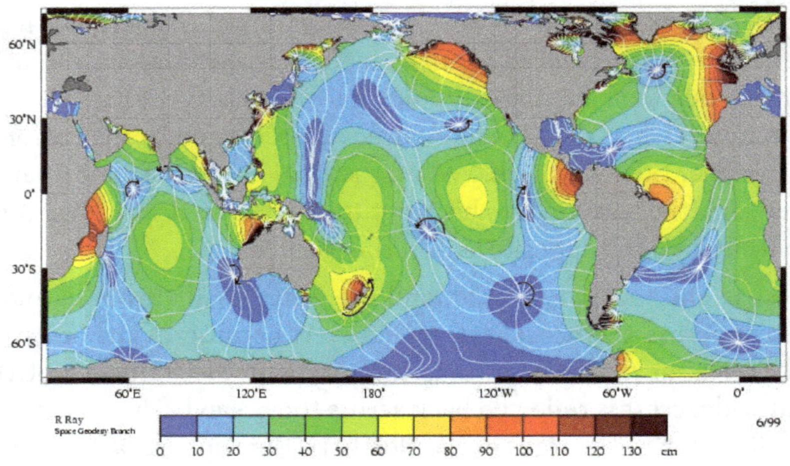

Fig. 5 Amphidromic circulation. Amphidromic points are represented where the white lines converge in areas of minimal tidal range. Tidal crests rotate around the amphidromic points, counterclockwise in the Northern Hemisphere and clockwise in the Southern Hemisphere.

Oceans and Some Rivers

There is a difference between having noticeable tides and having true tides. For tides to be noticeable, the body of water has to be huge, like an ocean. Even though true tides also occur in smaller water basins, like big lakes, the tidal variations here are too small to notice.

For example, in the Great Lakes in the US, the largest tidal range is less than 5 cm or just under 2 inches. Different weather conditions, such as wind and barometric pressure, creates bigger differences in the water level than tides on these lakes. This is also the case in the Baltic Sea, the Black Sea, the Caspian Sea, and even the Mediterranean.

Many rivers connecting to the ocean do have high and low tides. In some of these tidal rivers, the water drains away almost entirely at low tide, making it possible to walk across the bottom of the river. If a part of a larger river is affected by the tides, the section affected is known as tidal reach.

In a few areas, where the tide comes into a narrow bay or river, tidal bores can form. Created by the incoming tide, tidal bores are waves which travel against the direction of the water current.

Tides in the Human Body

Many people believe that the Moon's gravitational force also affects humans, as our bodies are made up of approximately 70% fluid. However, there is no scientific evidence supporting this belief. Compared to the Earth's oceans, the human body is far too small and contains far too little liquid to experience tides caused by the Moon in any meaningful way.

TIDAL PHENOMENA IN COASTAL REGIONS

Remember that the tides are fundamentally a wave. When tide waves enter coastal waters, they are subject to reflection and amplification similar to what wind-generated waves experience. In certain locations, reflected wave energy causes water to slosh around in a bay, producing standing waves. As a result, Large lakes and coastal rivers experience tidal phenomena, too. In some low-lying rivers, for instance, a tidal bore is produced by an incoming high tide. Further, the tides profoundly affect the behavior of certain marine organisms.

An Example of Tidal Extremes: The Bay of Fundy The largest tidal range in the world is found in Nova Scotia's Bay of Fundy.With a length of 258 kilometers (160 miles), the Bay of Fundy has a wide opening into the Atlantic Ocean. At its northern end, however, it splits into two narrow basins, Chignecto Bay and Minas Basin (Fig. 6).

The period of free oscillation in the bay—the oscillation that occurs when a body is displaced and then released—is very nearly that of the tidal period. The resulting constructive interference—along with the narrowing and shoaling of the bay to the north—causes a buildup of tidal energy in the northern end of the bay. In addition, the bay curves to the right, so the Coriolis effect in the Northern Hemisphere adds to the extreme tidal range.

During maximum spring tide conditions, the tidal range at the mouth of the bay (where it opens to the ocean) is only about 2 meters (6.6 feet). However, the tidal range increases progressively from the mouth of the bay northward. In the northern end of Minas Basin, the maximum spring tidal range is 17 meters (56 feet), which leaves boats high and dry during low tide (Fig. 6, insets).

COASTAL TIDAL CURRENTS

The current that accompanies the slowly turning tide crest in a Northern Hemisphere basin rotates counterclockwise, producing a rotary current in the open portion of the basin. Friction increases in nearshore shoaling waters, so the rotary current changes to an alternating or reversing current that moves into and out of restricted passages along a coast.

The velocity of rotary currents in the open ocean is usually well below 1 kilometer (0.6 mile) per hour. Reversing currents, however, can reach velocities up to 44 kilometers (28 miles) per hour in restricted channels such as between islands of coastal waters.

Reversing currents also exist in the mouths of bays (and some rivers) due to the daily flow of tides. Figure 7 shows that a flood current is produced when water rushes into a bay (or river) with an incoming high tide. Conversely, an ebb current is produced when water drains out of a bay (or river) because a low tide is approaching. No currents occur for several minutes during either high slack water (which occurs at the peak of each high tide) or during low slack water (at the peak of each low tide).

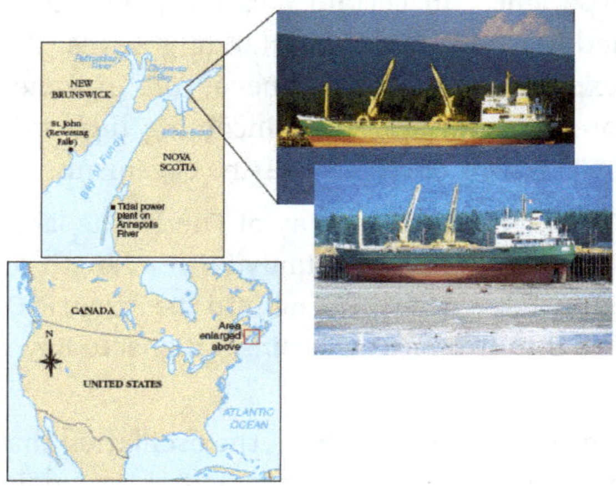

Fig. 6. The Bay of Fundy, site of the world's largest tidal range. Even though the maximum spring tidal range at the mouth of the Bay of Fundy is only 2 meters (6.6 feet), amplification of tidal energy causes a maximum tidal range at the northern end of Minas Basin of 17 meters (56 feet), often stranding ships (insets).

The analysis of tidal phenomena in narrow embayments

The tides of coastal embayments derive their energy from the ocean tides rather than from the direct action of lunar and solar gravitational forces. They are considered to be part of co-oscilating systems in which the period is determined by the tide in the outer sea, while the detailed character of the motion depends on the size and form of the enclosed basin . In narrow basins of simple form in which the influence of the earth's rotation is small, the motions resemble standing waves. Ideally, such waves are characterized by the simultaneous rise and fall of level on either side of a nodal line at which no change in elevation takes place. The elevation at high water increases with distance from the nodal line and slack water coincides with high and low water. The properties of tides due to standing waves may be deduced by assuming the motion to result from a primary progressive wave moving up the channel which undergoes complete reflection at a barrier.

Mathematically, this situation may be treated as the interference of two identical progressive waves moving in opposite directions and so related that both waves are in phase at the barrier. This treatment of standing waves assumes the presence of total reflection, the absence of damping and the absence of effects of the earth's rotation. Since these conditions are not realized in natural tidal basins, the standing wave concept leads to oversimplification. In coastal embayments the most striking departure from the expectations of the standing wave concept is the discrepancy between times of high water and slack water, which may be great near the mouths of the larger bays and sounds. High water does not occur simultaneously within such enclosures but is earlier near the sea. Commonly, the nodal line is represented merely by a region in which the tidal range is small. These are effects which can be explained if damping of the primary and reflected waves by frictional or other effects is taken into account.

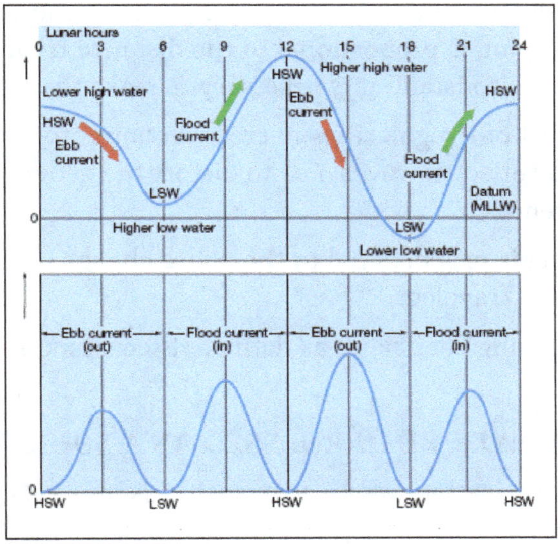

Fig. 7. Reversing tidal currents in a bay.
Top:Tidal curve for a bay, showing ebb currents are created by an outgoing low tide and flood currents are created by an incoming high tide. No currents occur during either high slack water (HSW) or low slack water (LSW).The datum MLLW stands for mean lower low water, which is the average of the lower of the two low tides that occur daily in a mixed tidal pattern. Bottom: Corresponding chart showing velocity of ebb and flood currents.

According to these concepts, the problem of tidal behavior in embayments is to determine numerically the properties of the primary and reflected waves so as to account for the observed relations of amplitude and stream velocity of the actual tide and to correlate these numerical properties with the geographical form of the embayment. In the present paper an attempt is made to treat the tidal behavior in such a way that the observed changes in elevation and motion of the water along the path of the wave may be used to determine the distribution of phase of the primary and reflected waves along the channel and to measure the damping.

The relations between the several aspects of a wave as it advances along a channel of uniform depth and width have been developed theoretically so as to show the times of high water and slack water, the range of the tide, and the phase relations of the primary and reflected waves along the channel for any degree of damping. By expressing the relationship of the several aspects of a reflected wave in a form in which the wave period is taken as the unit of time and distance is given in terms of the related phase changes, it is possible to eliminate the purely geographical dimensions and to obtain a wholly general description of the tide which may be used to indicate how any given channel distorts the behavior of the wave as it advances. In the case of irregular channels, in order to justify the application of relations deduced for uniform channels, in which the change in phase of the primary and reflected waves and their damping is proportional to the distance traveled and in which the velocity of the waves is constant, it is necessary to make the following assumptions:

1. That the effect of irregularities in cross section is to alter the velocity of the primary and reflected waves; i.e., to distort the geographical distribution of phase differences.
2. That damping is proportional to the phase change in the waves rather than to the distance traveled.
3. That the damping coeffcient, as defined, is constant along the length of the channel.

CAN TIDAL POWER BE HARNESSED AS A SOURCE OF ENERGY?

Tidal energy is a form of renewable energy that is generated by the tides. Tidal energy can be harnessed in two ways: by harnessing the kinetic energy of the tides or by harnessing the potential energy of the tides. Tidal energy is a reliable source of renewable electricity, and it has the potential to be a major source of renewable energy in the future.

Tidal energy is one such source of renewable energy. It also happens to be one of the oldest forms of renewable energy generation. The basic concept of tidal energy is the conversion of the kinetic or potential energy derived from tides into electricity.c Tidal energy is one such source of renewable energy. It also happens to be one of the oldest forms of renewable energy generation. The basic concept of tidal energy is the conversion of the kinetic or potential energy derived from tides into electricity.

Various setups used to harness energy from tides

1. **Tidal Barrages:** A tidal barrage is a dam that taps the potential energy generated by the difference in the height of the high tide and the low tide. The gates of the barrages are kept closed when it is high tide. This results in the water accumulating on one side of the dam. As the tide starts falling, the level of the water on the ocean side of the dam falls. When the difference in height is more than 5 meters, the dam is opened.

   This allows the water to flow through the turbines to the lower side. This sudden movement of water rotates the turbines, which generates electricity. The barrages can generate electricity about 4 times a day due to the two falling and two rising tides.

2. **Tidal Stream Systems:** These can be regarded as the windmills of the sea since, instead of air, they rely on water currents. The underwater turbines are anchored to the sea bed. These systems exploit the kinetic energy from the water to drive the blades of the rotor.

   The blades of these rotors are stronger and smaller than those of conventional windmills. This enables them to endure extreme pressure underwater. Tidal streams are gaining popularity, as they are cheaper to set up and maintain. The blades are equipped with a functionality that enables them to change direction according to the flow of current. This tidal stream technology is suitable in areas where the water currents are strong.

3. **Tidal Lagoons:** These are similar to barrages, except that lagoons are highly localized and specific to a particular stretch of water. Barrages span the entire width of a river or estuary. In the case of lagoons, the structure is confined to the parts with a high tidal range. Basically, lagoons are miniature barrages, the only difference being that a barrage blocks an entire estuary, whereas lagoons block only a certain part of the river.

   Although tidal energy is extremely reliable and clean, implementing it has its fair share of challenges, including high construction costs, disruption of local ecosystems and frequent required maintenance. That being said, it has a very high potential, with many advantages over solar and wind energy. Let's hope that the governments of the world make the most of this readily available natural resource in our ongoing efforts to be green and sustainable for generations to come.

HOW IS TIDAL POWER GENERATED?

All the coastal regions experience two high tides and two low tides for a period of 24 hours. In such cases, the tidal difference between low and high tides has to be harnessed into electricity. The difference must be more than 16 ft for generating electricity from tides. However, only around 40 regions on Earth have tidal ranges of this magnitude.

Tidal power can be generated using the following technologies:

Barrages or Dams

Barrages are built across a bay or river estuary. Each of them consists of gates for allowing the water to flow into the barrage along with the incoming tide. These gates are then closed for preventing the tides from going out. The water trapped inside the barrage is known as a hydrostatic head. More power can be generated with the large amount of head. When the tides produce an adequate difference in the water level, the other gates on the opposite side of the barrage is opened. The water then flows through the turbines, which turn the generators for producing electricity.

Tidal Fences

Tidal fences are giant turnstiles that have the potential to cross the straits between an island and the mainland and channels between small islands. The turnstiles can be rotated through tidal currents of coastal waters. These currents run at 5.6 to 9 mph for generating winds of higher velocity. As seawater has high density than air, ocean currents produce more energy than air currents.

Tidal Turbines

Tidal turbines are arranged underwater in rows like wind turbines. These turbines can work best in regions where the coastal currents run between 4 and 5.5 mph. With currents of that speed, a 15-m diameter tidal turbine is capable of producing energy equivalent to that of a 60-m diameter wind turbine. Tidal turbine farms can be located close to shore in water at a depth of 65.5 to 98.5 ft.

Advantages of Tidal Power

The key benefits of tidal power include the following:

- Tides are easily predictable
- Inexpensive to maintain
- Reliable and renewable source of energy
- High energy density than other renewable energy forms
- It produces no greenhouse gases or other waste
- Vertical-axis turbines and offshore turbines are inexpensive to build and have less environmental impact
- Tidal turbines are 80% efficient, which is higher than solar or wind energy generators.

- Barrages reduce the damage of high tidal surges on the land.

Disadvantages of Tidal Power

The following are some of the limitations of tidal power:

- Initial construction cost is very high
- Formation of silt behind the barrage
- Effect on animals and plants living near tidal stations
- Very few suitable sites for constructing barrages
- Disturbs migration of living creatures in the ocean
- Water cannot be replenished, and hence dirt gets settled within the coast
- It produces power for only about 10 h of the day when the tide is moving in and out.

Environmental Aspects of Tidal Power

Several assessments have identified a number of potential environmental impacts of tidal energy development. Some of them are listed below:

- Emission of electromagnetic fields
- Interference with animal migration and movements
- Noise during construction and operation
- Alteration of waves and currents
- Alteration of substrates, sediment transport and deposition
- Change in habitats of benthic organisms

Conclusion

According to researchers, 2013 could experience a big breakthrough in tidal technologies. MeyGen, a tidal technology company, is planning to use tidal technology in the Pentland Firth that can generate up to 40MW of electricity, sufficient for nearly 38,000 homes during its initial phase. Researchers feel that tidal technology development is more reliable than wind and the quality of the power produced by tidal sources is better than other renewable resources. However, the high capital costs associated with the development of tidal technologies are likely to restrict the development of tidal power in the near future.

CHAPTER 10
BEACHES, SHORELINE PROCESSES AND THE COASTAL OCEAN

BEACHES

Beaches are wave-deposited accumulations of sediment located at the shoreline. They require a base to reside on, usually the bedrock geology, waves to shape them, sediment to form them, and most are also affected by tides. The beach extends from wave base where waves begin to feel bottom and shoal, across the nearshore zone, though the surf zone to the upper limit of wave swash (Fig. 1). In the coastal zone ocean waves are transformed by shoaling, breaking, and swash. In doing so they interact with the seabed, and determine the beach morphology or shape, a process called beach morphodynamics.

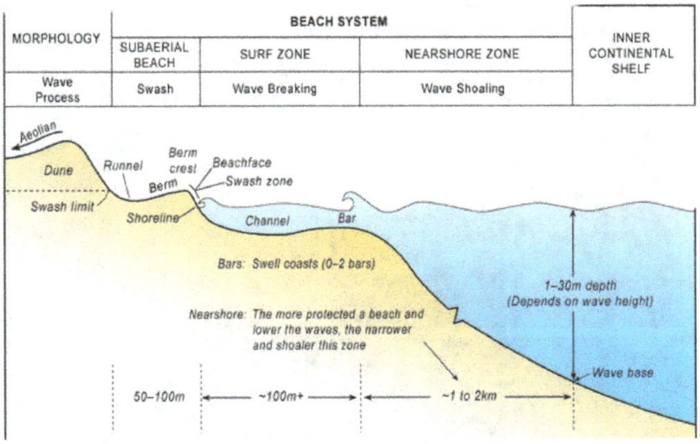

Fig. 1. An idealised cross-section of a wave-dominated beach system consisting of the swash zone which contains the subaerial or 'dry' beach (runnel, berm, and beach face) and is dominated by swash processes; the energetic surf zone (bars and channels) with its breaking waves and surf zone currents; and the nearshore zone extending out to wave base where waves shoal building a concave upward slope.

For most people, when they think of coastal areas they picture a beach, and the beach that they imagine is probably a typical sandy beach composed of quartz sand grains. But beaches are comprised of whatever types of sediments are dominant in the local area. For example, parts of Hawaii and Iceland are famous for their black sand beaches, made up of eroded basalt and other volcanic materials.

The beautiful tropical white sand beaches we see in travel ads are largely composed of the crushed calcium carbonate remains of coral skeletons (much of which has been chewed up and excreted by a fish before we happily run our toes through it!) Other beaches may lack sand altogether and instead be dominated by small shells, or larger rocks or pebbles (Fig. 2).

The shoreline is divided up into multiple zones (Fig. 3). The backshore is the region of the beach above the high tide line, which is only submerged under unusually high wave conditions, such as during storms. The foreshore lies between the high tide and low tide lines; it is submerged during high tide and is exposed during low tide. The nearshore extends from the low tide line to the depth where wave action is no longer influenced by the bottom, i.e. to where the depth exceeds the wave base. Finally, the offshore zone represents the depths beyond the nearshore region.

Along the beach itself, the area above the high tide line is called the berm, which is usually dry and relatively flat. The berm often ends with a berm crest or berm scarp, which is a steeper wall carved out by wave action that leads down to the foreshore. The foreshore has a number of other names, including the beach face, the intertidal or littoral zone, and if the area is fairly flat, the low tide terrace. Just off shore from the beach there are often longshore bars and longshore troughs running parallel to the beach. The longshore bars are accumulations of sand that are deposited by wave action and longshore currents. The decrease in depth above longshore bars is what often causes waves to start to break well before reaching the beach.

Fig. 2 Various beach substrates. Clockwise from top left: Punaluu Black Sand Beach, Hawaii, USA

The sand or other particles that make up the beach are distributed by wave action. The water that moves over a beach through incoming waves is called swash, and it also contains suspended sand grains that can get deposited on the beach. Some of the swash percolates into the sand while the rest of the water washes back out as backwash as the wave recedes. Backwash removes sand from the beach and returns it to the ocean.

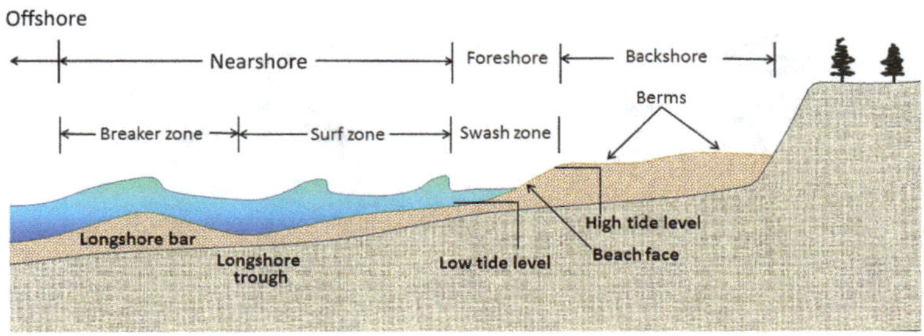

Fig. 3 The zones of a typical beach

Sand will therefore be deposited or eroded depending on which process is dominant. If wave action is light, a lot of incoming water gets absorbed by the sand, so swash dominates. Under heavier waves the beach becomes saturated with water, so less can be absorbed, and backwash is dominant. This leads to seasonal cycles in beach structure; waves are heavier during the winter as a result of stormier conditions at sea, so backwash dominates and sand is removed from the beach and deposited offshore in longshore bars. In the summer the waves are gentler, swash dominates, and the sand is transported from the longshore bar and deposited on the shore to create a wider, sandy beach (Fig. 4).

Beach sediment - types & sources

All beaches consist of sediment, which can range in size from sand up to cobbles and boulders. The finer sand result in very low gradient (~1°) beaches while cobbles may be stacked as steep as 20° (Fig. 5a and 5b). Most beaches with fine to medium sand have a swash zone gradient between 1-8°.

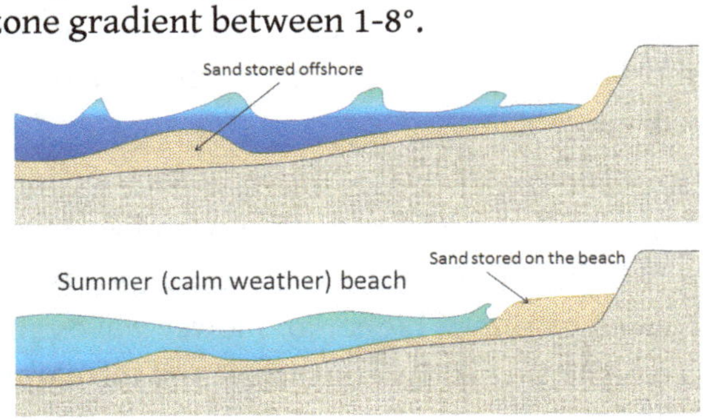

Fig. 4 The differences between summer and winter on beaches in areas where the winter conditions are rougher and waves have a shorter wavelength but higher energy. In winter, sand from the beach is stored offshore

In the mid latitudes most beaches are composed of siliceous or quartz sand grains derived from erosion. In the topics, coral reef detritus and shells known as 'carbonate sediment' tend to dominate, (Fig. 5c and 5d), while in higher latitudes physical weathering produces coarse rock fragments and gravel.

Fig. 5. a) A steep cobble-boulder beach at Catfish Bay, Western Australia. (b) Low-gradient fine sand beach at Ball Bay, Queensland. (c) Shell Beach, Western Australia is composed entirely of the white cockle shell *Fragum erugatum*. (d) Close up view of the cockle shell.

Beach sub-systems

At the beach the three zones of wave transformation (shoaling, breaking, and swash) produce three morphologically distinct sub-systems. The shoaling wave zone builds a low gradient (~1°) concave upward profile, smooth in outline, with small wave-generated ripples and generally onshore sediment movement. As they shoal they interact with the seabed, slowing down and increasing in steepness and heigh.

The surf zone is the most dynamic part of the beach and extends from the breaker zone to the shore. Waves break when the water depth is approximately 1.5 times the wave height. They can break as a spilling breaker on low gradient slopes, a plunging wave on moderate gradients, or a surging wave on steep slopes. In breaking, waves transform their potential energy to kinetic energy, which is initially manifest as the broken wave of translation, or wave bore, which moves shoreward as broken white water. At the shoreline the currents can be deflected longshore and water may return seaward as a rip current.

Surf zone currents can transport sediment onshore, longshore and offshore and build the (sand) bars and troughs that occupy the surf zone. The number and location of the bars is a product of infragravity waves, a low frequency (greater 30 sec period) wave produced by sets of higher and lower waves and which is enhanced by wave breaking across the surf zone. The longer the infragravity wave period the more widely spaced the bar(s).

Another form of infragravity wave called edge waves also influence the longshore spacing of rip currents and channels, which are typically 200-300 m apart on ocean beaches. Rip currents are narrow, seaward moving currents that move seaward though the surf zone, often in a deeper rip channel. They are a mechanism for returning the water back out to sea, and a conduit to transport seaward eroded beach sediment during high seas. They are also a major hazard to beach goers and responsible for most beach rescues and drowning.

When the broken wave reaches the base of the wet beach it collapses and runs up the beach face as swash or uprush in the swash zone. The uprush stops toward the top of the slope, some percolates into the beach, the remainder flows back down the beach as backwash. Both actions produce a relatively steep seaward sloping swash zone or beach face, which can range from 1 to 20°. As sediment is deposited in the swash zone it can build a berm, a near horizontal to slightly landward-dipping sand surface, the area where most people sit when they go to the beach. The swash zone may also contain beach cusps, spaced about every 20 to 30 m and produced by another form of edge wave.

SHORELINE PROCESSES

Another important effect of waves reaching the shore at an angle is that when they wash up onto the beach, they do so at an angle, but when that same wave water washes back down the beach, it moves straight down the slope of the beach (Fig. 6). The upward-moving water, known as the swash, pushes sediment particles along the beach, while the downward-moving water, the backwash, brings them straight back. With every wave that washes up and then down the beach, particles of sediment are moved along the beach in a zigzag pattern.

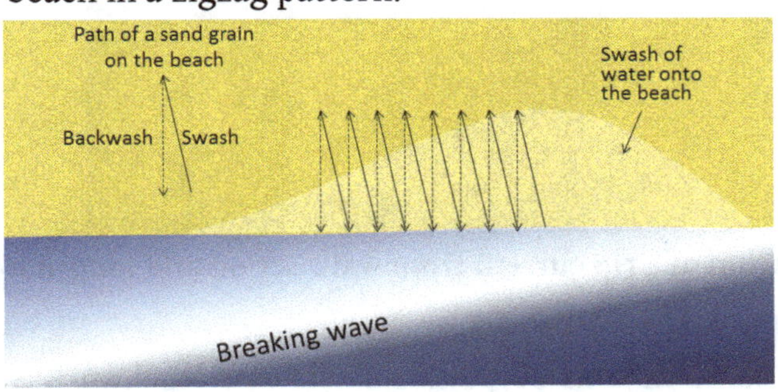

Fig. 6 The zigzag pattern of sediment movement along a beach creating longshore transport. In this figure the longshore transport moves particles to the left

The combined effects of sediment transport within the surf zone by the longshore current and sediment movement along the beach by swash and backwash is known as longshore transport, or littoral drift. Longshore transport moves a tremendous amount of sediment along coasts (both oceans and large lakes) around the world. The net movement of sediment due to longshore transport is to the south along both coasts of the continental United States, because the storms and high winds that originally create the swell tend to occur at higher latitudes and move to the south.

A rip current (often incorrectly called a "rip tide"; they are not really related to tides) is another type of current that develops in the nearshore area, and has the effect of returning water that has been pushed up to the shore by incoming waves or accumulated through longshore currents, particularly converging longshore currents. Rip currents often occur where there is a channel between sandbars that makes it easier for the retreating water to escape.

As shown in Fig. 7, rip currents flow straight out from the shore, and because the water is directed through a narrow space, the current can be very strong. The currents lose strength quickly just outside of the surf zone, but they can be dangerous to swimmers who get caught in them and are pulled away from shore. Swimmers caught in a rip current should not try to swim directly back to shore, as it is difficult to fight the current and the swimmer can quickly tire. Instead, swim parallel to the beach for a short distance until you are outside of the rip current, and then you can easily swim to shore.

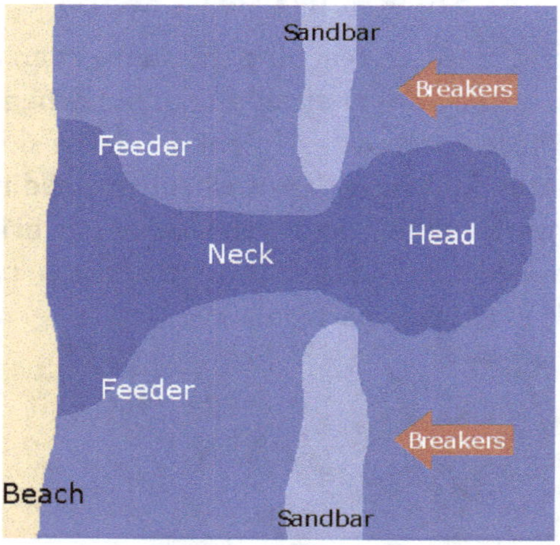

Fig. 7 Creation of a rip current from wave action and longshore transport.

Longshore current

- Happens same time as perpendicular (swash/backwash induced) sediment transport.
- Wave refraction causes water and sand to move parallel to shore
- Zigzag motion in surf zone
- Longshore current
- Longshore transport
- Refraction causes wave to bend and break nearly parallel to shore
- Wave strike (the swash) at a small angle

- The backwash is straight down the beach face
- Results in zig-zag sediment movement, whose net transport is along shore

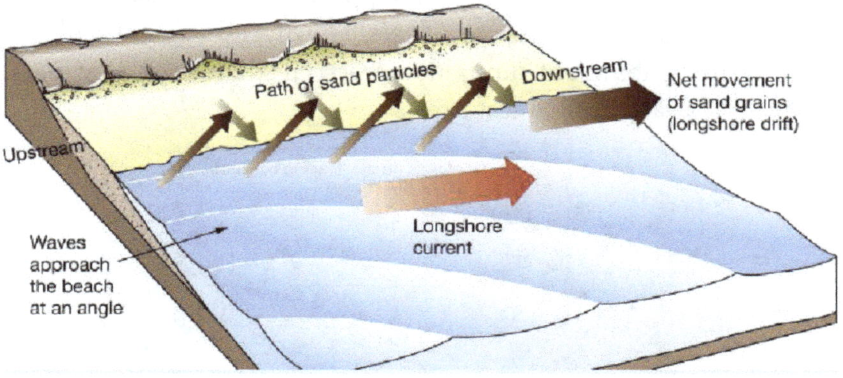

Fig. 8. Longshore current

Erosional shorelines

- Well-developed cliffs
- Recent tectonic activity
- Headlands
- Wave-cut cliff with sea cave
- Some flattening of the near shore bedrock (the wave cut bench)
- Sea arches
- Sea stacks
- Marine terrace
- The wave cut bench that gets uplifted ? Wave erosion increases with:
- More shore exposed to open ocean
- Smaller tidal range
- The wave energy is more focused on a narrow strip of cliff rather than distributed across a large area
- Weaker bedrock

COASTAL PROCESSES

The coastal zone is that part of the land surface influenced by marine processes. It extends from the landward limit of tides, waves, and wind blown coastal dunes, and seaward to the point at which waves interact significantly with the seabed.The coastal zone is a dynamic part of the Earth's surface where both marine and atmospheric processes produce rocky coasts, as well as beaches and dunes, barriers and tidal inlets, and shape deltas. The atmospheric processes include temperature, precipitation, and winds, while the major marine processes are waves and tides, together with water temperature and salinity. The coast also supports rich ecosystems, including salt marshes, mangroves, seagrass, and coral reefs. The diverse coastal ecology is favored by the shallow waters, abundant sunlight, terrestrial and marine nutrients, tidal and wave flushing, and a range of habitat types.

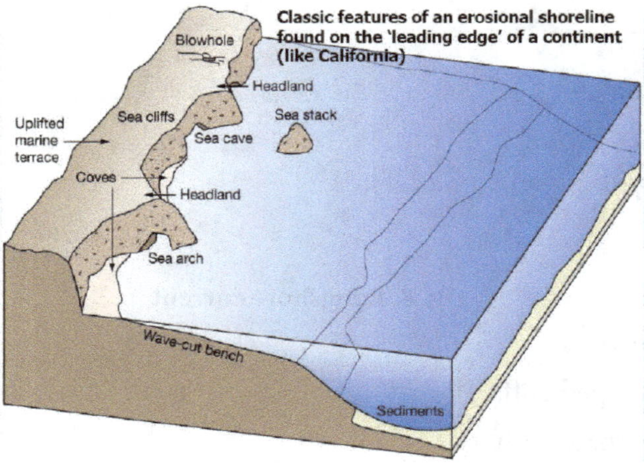

Fig. 9. Erosional shorelines

Waves - generation and types

Waves provide about half the energy to do work at the coast. Ocean waves are generated by wind blowing over the ocean surface. The stronger the wind, the longer it blows and the longer the fetch, or stretch of ocean over which it blows, the larger the waves (Fig. 10).

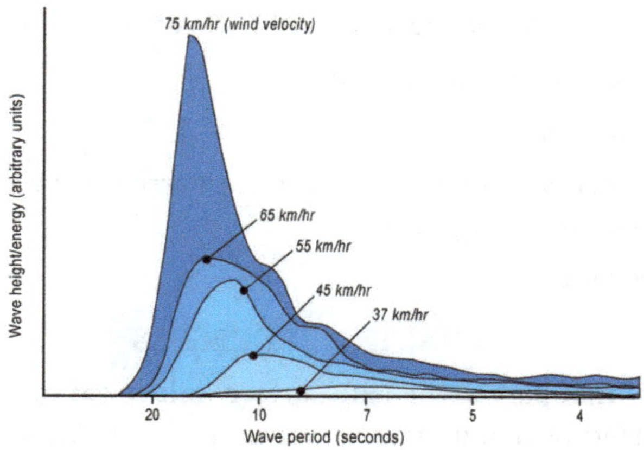

Fig. 10. As wind velocity increases, the period or time between waves, and wave length, increases, and the amount of energy transferred to the waves increases exponentially. Note how as wind velocity doubles from 37 to 75 km/hr the amount of energy increases exponentially. Very strong winds are therefore required to generate the biggest waves.

The world's greatest wave factories are in the zone of sub-polar lows centered on 40-60° N and S latitudes, the so-called roaring 40's and screaming 60's. The strong westerly winds produce the world's biggest waves which initially head west, and are deflected equatorward by the Coriolis effect, arriving from the northwest in the northern hemisphere and southwest in the southern hemisphere (Fig. 11).

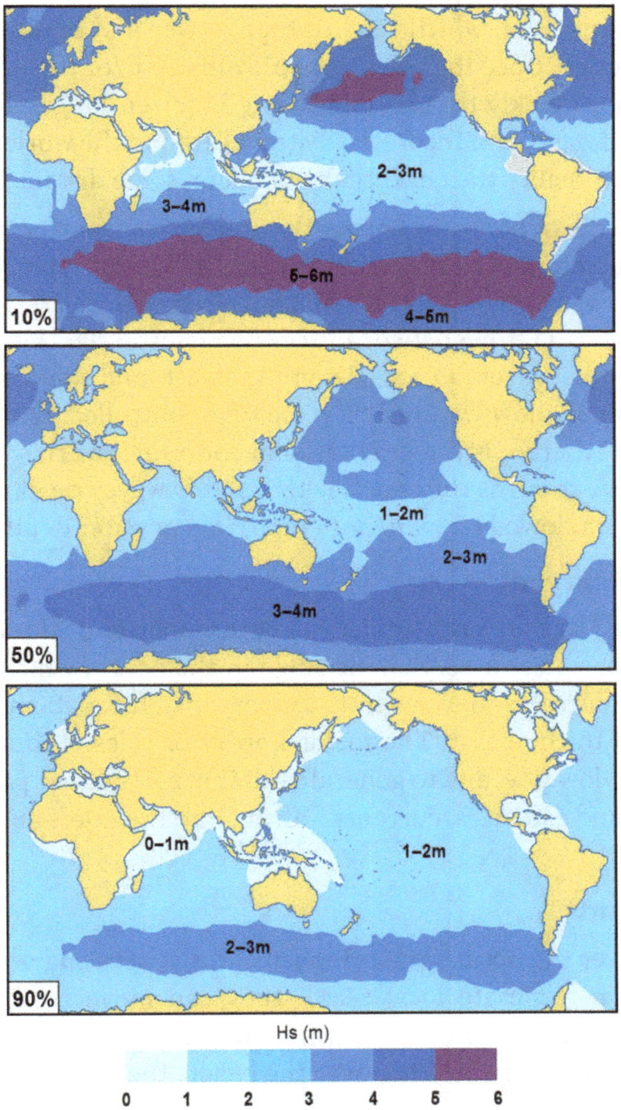

Fig. 11. The highest waves occur in the Southern Ocean and north Pacific and Atlantic where they are generated by strong sub-polar lows. They average 5-6 m in height 10% of the year (top). The Southern Ocean has the most persistent higher waves averaging 3-4 m 50% of the time (centre), and at least 2-3 m 90% of the year (bottom). While the sub-tropical trade winds also blow year round across huge sections of ocean, they are of only moderate velocity, and hence generate only moderate waves.

Other major wave climates are the easterly waves produced by the expansive but moderate velocity northeast and southeast Trade winds - and lesser seasonal waves produced by the monsoons and even the polar easterlies, together with occasional hurricanes that can produce massive waves as well as storm surges.

When waves are being generated they are called a sea and consist of short, steep, high, slower waves, which tend to topple over and break, and have a broad spectrum of direction. Once the wind stops blowing, and/or the waves leave the area of generation, they quickly transform into swell - lower, longer, faster and uniform in direction. Swell waves can theoretically travel around the world with minimal loss of energy, while in reality they eventually break on some distant shore.

Waves are a form of potential energy that can be transported across hundreds to thousands of kilometers of ocean to be released as kinetic energy when they shoal and break. Waves are defined by their height (H) (trough to crest), length (L) (crest to crest) and period (T) (time between successive crests). The longer the period the longer and faster the waves, as wave length $L=1.56\ T$, and wave velocity $C = 1.56\ T^2$. When waves enter shallow water their velocity is controlled by the water depth (d) such that $C = \sqrt{gd}$ where g is the gravitational constant. For this reason waves slow down as they move towards the shore, with a 10 sec wave traveling at 56 km/hr (35 mph) in deep water slowed to 7 km/hr (4 mph) in 5 m water depth.

Tides and tidal currents

Tides are produced by the gravitational pull of the Moon and Sun acting on a rotating Earth. This pull produces a very slight bulge in the ocean, which we know as tide. The tides and the currents they generate are responsible for about 50% of the marine energy delivered to the coast. The major impact of tides is to shift the shoreline between high and low tide, and to generate tidal currents either parallel to the coast, or at tidal inlets and estuaries, currents flowing into the inlets and perpendicular to the coast.

Wind and currents

Winds blowing over the oceans are responsible for generating ocean waves. Nearer the coast they can generate local seas - they can move the ocean surface and generate locally wind driven currents which in places can result in upwelling and downwelling. Finally, when blowing over the beach, they can transport sand inland to build coastal sand dunes.

Fluvial-deltaic systems

Fluvial systems deliver sediment to the coast where it is deposited in estuaries and deltas. Depending on their location, deltas are also acted on by waves, tides, and other currents, and shaped to suit the prevailing processes. Sediment can also be moved longshore to supply beach and barrier systems (Fig. 12).

Sea level

Sea level determines the position of the shoreline. During the last glacial maxima (ice age) 18,000 years ago, sea level was 120 m below present, and the continental shelves were exposed. It then rose, reaching present sea level around 6,000 years ago, after which it was relatively stable. Now, with climate change, it is beginning to rise again, and may rise as much as 1 m over the next 100 years, triggering shoreline retreat, inundation, and erosion.

Fig. 12. The Gasgoyne River delta in Western Australia delivers large volumes of sand to the coast where it is deposited in river mouth shoals and slowly reworked longshore to supply downdrift spits, barriers, and dunes.

SAND MOVEMENT

There is a constant flow of sand from the land into the ocean. Watershed run-off and bluff and hillside erosion bring sand to the beach. Sand grains travel southward down the coast, while finer particles of sediment are carried and deposited further out to sea.

Fig. 13. Sand Movement

Along the way, sand is washed ashore, temporarily resting on beaches, until it is re-suspended in the ocean by wave action or wind. The one-way journey down the coast ends when sand is blown inland forming sand dunes, or more commonly, when it flows into a submarine canyon. This deep underwater feature is essentially the dead end of a littoral cell, where sand is deposited for the long-term and, for practical purposes, lost.

Littoral Cell

A littoral cell is a distinct area of the coastline where sand enters the ocean, flows down the coast, and then is removed from the system. Permanent loss of sand occurs at the end of the littoral cell when it flows into a submarine canyon or, less frequently, when it accumulates on shore as part of a sand dune. The amount of sand available to beaches is the amount of sand flowing into the littoral cell minus the amount flowing out. If this sand budget is altered, beaches can narrow or even disappear.

HOW DOES HARD STABILIZATION AFFECT COASTLINES?

Beaches and shorelines are dynamic places that are in a constant state of change due to forces that act upon them, such as waves and currents. Coastal erosion is constantly happening because waves are constantly hitting the beach and shoreline structures. However, this process can be accelerated by storms, which produce large waves that crash onto shore with a lot of energy. Erosion of beaches and shorelines can negatively impact the lives of people living along the coast and cause costly damage to their properties.

Therefore, coastal communities are often willing to invest in hard stabilization, which is the use of man-made protective structures to control erosion. While hard stabilization structures help slow erosion caused by waves, they also prevent the shifting of sand along the beach. Sand and sediment are constantly moving parallel to the shoreline due to the force of waves that strike the shore at an angle. Waves rarely strike the shore straight on. Instead, we see a phenomenon called wave refraction, in which the waves bend as they travel toward the shallow waters of the shore.

These oblique waves create the longshore drift, which is the process by which sand and sediment are transported along the coast. The longshore drift can be responsible for the loss of beach sand from an area, and you can imagine that people who own homes, restaurants or hotels along that stretch of beach suffer from this loss. They do not want to watch the beach in front of their property deteriorate as sand is washed down the coast.

Shoreline hardening, or hard stabilization, involves construction of hard immovable engineered structures, such as seawalls, rock revetments, jetties, and groins. Seawalls and rock revetments run parallel to the beach. Seawalls are vertical structures, constructed parallel to the ocean shoreline, and are primarily designed to prevent erosion and other damage due to wave action. Revetments are shoreline structures constructed parallel to the shoreline and generally sloped in such a way as to mimic the natural slope of the shoreline profile and dissipate wave energy as the wave is directed up the slope. Breakwaters are structures constructed waterward of, and usually parallel to, the shoreline.

They attempt to break incoming waves before they reach the shoreline, or a facility (e.g., marina) being protected. Jetties and groins are manmade structures constructed perpendicular to the beach, with jetties usually being much longer, and are located adjacent to inlets with the purpose of maintaining navigation in the inlet by preventing sand from entering it.

In contrast, terminal groins are structures built at the end of a littoral cell to trap and conserve sand along the end of the barrier island, stabilize inlet migration, and widen a portion of the updrift beach. Terminal groins are designed so that when the area behind the groin fills in with sand, additional sand will go around the structure and enter the inlet system.

It is well accepted that hard stabilization techniques along high energy ocean shorelines will accelerate erosion in some location along the shore as a result of the longshore sediment transport being altered). The hydromodifications resulting from coastal armoring modifies sediment grain size, increases turbidity in the surf zone, narrows and steepens beaches, and results in reduced intertidal habitat and diversity and abundance of macroinvertebrates. A study looking at the effect of a short groin (95m) on the benthic community found that the groin created a depositional condition on one side of the structure and erosion on the other, and macroinvertebrate diversity and abundance was significantly reduced within 30m of the structure, as sand particle size and steepness increased.

The change in benthic community was attributed to the change in geomorphology of the beach. Hard structures along a sandy beach can also result in establishment of invasive epibenthic organisms. A secondary impact of hardened structures is that the areal loss of beach resulting from hardening of shorelines is often managed by implementing nourishment projects, possibly having additional damage to subtidal bottom. Anchoring inlets also prevents shoal formation and diminishes ebb tidal deltas, which are important foraging grounds for many fish species. Recognizing that hardened structures are damaging to recreational beaches and the intertidal zone, four states have prohibited shoreline armoring: Maine, Rhode Island, South Carolina, and North Carolina.

Perhaps the greatest impact of terminal groins and jetties results in the long-term effect on barrier islands and the effect that will have on marine and estuarine ecosystems. By stabilizing the inlet, inlet migration and overwash processes are interrupted, causing a cascade of other effects.

In the case of Oregon Inlet, the terminal groin anchored the bridge to Pea Island and stopped the migration of the inlet on the south side. But the continuing migration of the north end of Bodie Island led to an increased need for inlet dredging. The combination of reduced longshore transport of sediment due to the groin and the post-storm dune construction to open and protect the highway prevented overwash processes that allow Pea Island to maintain its elevation over time. With overwash processes disrupted, the beach profile has steepened, and the island has flattened and narrowed, increasing vulnerability to storm damage.

At Oregon Inlet and Pea Island, the accelerated need for beach replenishment is further aggravated by the need to maintain Hwy 12 on the narrowing beach. From 1983 to 2009 approximately 12.7 million cubic yards of sand have been added to the shoreline within three miles of the terminal groin. Researcher documented that the large volumes of sand replenishment in this area, required to maintain the channel, protect the road, and maintain a beach have resulted in a significant reduction in grain size and reduction in mole crab abundance.

Mole crabs are considered an important indicator of beach conditions due to their importance in the food web as prey for shorebirds and surf fish. In addition to causing erosion on downdrift beaches, altering barrier island migration processes, and accelerating the need for beach nourishment projects, jetties obstruct larval fish passage through adjacent inlets.

Disadvantages of Hard Structures

While hard coastal structures can be the most effective option for flood protection and/or mitigation, or for stabilizing a shoreline at a fixed position, there is a price to pay. Hard structures partially hinder the recreational use of the coastal zone and can cause adverse ecological effects within the coastal zone. For example, when seawalls are constructed on eroding beaches, the erosion continues so that the beach in front of the seawall can become very narrow or disappear completely. And while groins and jetties trap sediment on the updrift side resulting in shoreline accretion, there is corresponding shoreline erosion on the downdrift side due to the interruption in longshore transport. Some of the disadvantages of hard structures include:

- Visual impacts
- Horizontal and vertical access restrictions
- Loss of sand supply to beach from armoring backshore
- Placement losses with construction of revetment or seawall
- Passive erosion
- Active erosion

COASTAL WATERS

As the interface between terrestrial environments and open oceans, coastal waters encompass many unique habitats and serve important human needs. Coastal habitats include estuaries, coastal wetlands, seagrass meadows, coral reefs, mangrove forests, kelp forests, and upwelling areas.

Coastal waters support many fish species and provide breeding habitat for 85 percent of U.S. migratory birds. They also provide habitat for many other organisms such as marine mammals, corals, sea turtles, and submerged aquatic vegetation.

Coastal waters support a wide range of human activities such as tourism, recreation, transportation, and fisheries. Lands on the coast are highly desirable places for people to live.

Coastal Water Extent

The extent of coastal waters - their spatial area - is particularly important with respect to specific types of coastal waters, such as coastal wetlands and coral reefs. Extent can be affected by natural factors, such as erosion and storms. It can also be changed by human activities, such as draining or filling wetlands for development or constructing seawalls or barriers. Channeling rivers can cause land to "sink" in coastal areas that otherwise would be naturally replenished by sediments.

Coastal Water Condition

The *condition* of coastal waters reflects a group of interrelated physical, chemical, biological, and ecological characteristics. Coastal water condition can be affected by a variety of stressors.

- Nutrients and pathogens can come from storm water, agricultural runoff, and sewage discharge or overflows. Excess nutrients can cause algal blooms that result in low dissolved oxygen levels, which harm aquatic life. Pathogens (e.g., bacteria and viruses) can affect the health of people who use waters for recreation or eat contaminated fish or shellfish.
- Chemical contaminants can come from sources such as agricultural runoff, industrial activities, and atmospheric deposition of airborne pollutants. Of particular concern to human health are toxic chemicals in consumable fish and shellfish.
- Changes in temperature and salinity can be influenced by weather patterns or the condition of freshwater inputs. These changes can affect habitat quality and the status of native plant and animal populations, and can also influence algal blooms.

- Non-indigenous species can affect the status of native communities. In particular, invasive species can kill or crowd out native populations or otherwise alter coastal watersheds.
- Overharvesting can affect populations of fish, shellfish, marine mammals, and other species.
- Changes in the extent of coastal waters can also affect their condition. For example, beach erosion and wetland loss can affect contaminant and sediment levels. Wetland loss can also affect the condition of the wetlands that remain.

COASTAL ECOSYSTEMS AND CLIMATE CHANGE

Coastal ecosystems are the unique habitats formed by plants and other organisms that can thrive at the borders between ocean and land, where they must live in saltwater and changing tides.

Like forests, many of these coastal ecosystems are full of plants that help regulate the Earth's temperature. As the plants in these ecosystems grow, they pull carbon out of the air and store it in their tissue, roots, and the soil beneath them. This keeps carbon out of our atmosphere, where, as the greenhouse gas carbon dioxide, it would otherwise trap heat and warm the planet.

Types of coastal ecosystems

Mangrove forests are a type of coastal forest made by mangrove trees, which have waxy leaves and characteristic roots that stick a few feet out of the water. These trees, which survive by filtering the salt out of seawater, form a dense forest along shorelines and tidal rivers. Mangrove forests are found in tropical and subtropical parts of the world, where they provide commercial fishing ground, timber and other economic goods for coastal communities.

Salt marshes are coastal wetlands that are flooded as the tides move in and out. Found mostly in temperate areas and higher latitudes, they are home to salt-tolerant grasses and bushes. Salt marshes provide shelter and breeding grounds for many kinds of fish and shellfish, as well as a resting place for migrating birds.

Seagrass meadows are formed by seagrasses, flowering underwater plants with long, thin leaves. They form a dense mat in shallow waters, with their roots anchoring the sediment beneath them in place. These meadows, which are found all the way from the tropics up to the Arctic Circle, are home to a diverse array of fish and other marine wildlife.

Other coastal ecosystems, like oyster reefs and coral reefs, are formed by animals instead of plants. Like plant-based coastal ecosystems, these reefs help protect the land from powerful storms.

Carbon storage and coastal protection

Coastal ecosystems are especially good at storing carbon because of the thick, rich layers of soil they build up. In most forests, when trees die, they release their carbon back into the atmosphere. But in many coastal ecosystems, when plants die, they don't fully break down, and their carbon can stay trapped in water-logged soils for thousands of years. In fact, mangroves store three to five times as much carbon per acre as other tropical forests. Coastal wetlands can also store carbon that comes from other ecosystems, by filtering out carbon-rich sediments suspended in river water as it flows out to sea. Because coastal ecosystems are such potent "carbon sinks," the loss of these habitats is an important driver of climate change.

Intact, these coastal habitats are actually one of our best ways to protect coastal communities from climate change. During hurricanes and other storms, high winds can push walls of water toward shore. Coastal habitats absorb energy from incoming waves and slow down these storm surges. As sea levels rise and hurricanes become stronger, we will need this protection more than ever.

Habitats under threat

While coastal wetlands are a key defense against climate change, these ecosystems are themselves vulnerable to climate change. In theory, coastal wetlands, whose plants can only survive in specific amounts of saltwater, could adapt to rising sea levels by moving inland. But if they're already on the edge of a steep slope, or a developed area like a road, they have nowhere to go. And in some areas, the rate of sea level rise could outpace wetlands' ability to build up enough soil to remain above water. More intense storms can also erode the soft soils of salt marshes and seagrass meadows.

These ecosystems face other, more direct threats from humans. As our cities expand, we fill in coastal wetlands to make room for development and farmland. In the tropics, fish farmers cut down mangrove forests for aquaculture ponds. We also harm these ecosystems by dredging deep channels for ships to navigate, and by polluting water with fertilizer, emissions from cars and buses, and other chemicals. And this further feeds climate change: when we destroy coastal ecosystems, their stored carbon gets released into the atmosphere as planet-warming carbon dioxide.

Protecting coastal ecosystems

There is a growing movement to preserve these habitats because of all the benefits they provide us. Conservation groups and others are protecting undeveloped land behind coastal ecosystems to help them adapt to sea level rise. The low-lying Netherlands has started shoring up some of the salt marshes in front of its dykes, to protect this infrastructure, and its citizens, from sea level rise. Some environmental policymakers have also proposed paying coastal landowners for the carbon stored in these ecosystems to prevent their destruction.

CHAPTER

MARINE POLLUTION 11

INTRODUCTION

Marine Pollution refers to trash and pollutants that come from land sources to end up in the ocean. This pollution causes widespread damage to ocean life as well as to economic structures that rely on marine infrastructure.

Marine Pollution happens when entry of industrial, agricultural, residential wastes etc enter the ocean. Air pollution is also a contributing factor when nitrogen, silicon, sulphur or pesticides are blown into the ocean.

Although it's land pollution that contributes 80% of marine pollutants, both the air and land variants are the primary contributors to marine pollution.

The majority of pollutants that make their way into the ocean come from human activities along the coastlines and far inland. One of the biggest sources of pollution is nonpoint source pollution, which occurs as a result of runoff. Nonpoint source pollution can come from many sources, like septic tanks, vehicles, farms, livestock ranches, and timber harvest areas. Pollution that comes from a single source, like an oil or chemical spill, is known as point source pollution. Point source pollution events often have large impacts, but fortunately, they occur less often. Discharge from faulty or damaged factories or water treatment systems is also considered point source pollution.

Eighty percent of pollution to the marine environment comes from the land. One of the biggest sources is called nonpoint source pollution, which occurs as a result of runoff. Nonpoint source pollution includes many small sources, like septic tanks, cars, trucks, and boats, plus larger sources, such as farms, ranches, and forest areas. Millions of motor vehicle engines drop small amounts of oil each day onto roads and parking lots. Much of this, too, makes its way to the sea.

Some water pollution actually starts as air pollution, which settles into waterways and oceans. Dirt can be a pollutant. Top soil or silt from fields or construction sites can run off into waterways, harming fish and wildlife habitats.

Non-point source pollution can make river and ocean water unsafe for humans and wildlife. In some areas, this pollution is so bad that it causes beaches to be closed after rainstorms.

More than one-third of the shellfish-growing waters of the United States are adversely affected by coastal pollution.

Correcting the harmful effects of nonpoint source pollution is costly. Each year, millions of dollars are spent to restore and protect areas damaged or endangered by nonpoint source pollutants. NOAA works with the U.S. Environmental Protection Agency, Department of Agriculture, and other federal and state agencies to develop ways to control nonpoint source pollution. These agencies work together to monitor, assess, and limit nonpoint source pollution that may result naturally and by human actions.

MARINE DEBRIS

Marine debris is a persistent pollution problem that reaches throughout the entire ocean and Great Lakes. Our ocean and waterways are polluted with a wide variety of marine debris, ranging from tiny microplastics, smaller than 5 mm, to derelict fishing gear and abandoned vessels. Worldwide, hundreds of marine species have been negatively impacted by marine debris, which can harm or kill an animal when it is ingested or they become entangled, and can threaten the habitats they depend on. Marine debris can also interfere with navigation safety and potentially pose a threat to human health.

All marine debris comes from people with a majority of it originating on land and entering the ocean and Great Lakes through littering, poor waste management practices, storm water discharge, and extreme natural events such as tsunamis and hurricanes. Some debris, such as derelict fishing gear, can also come from ocean-based sources. This lost or abandoned gear is a major problem because it can continue to capture and kill wildlife, damage sensitive habitats, and even compete with and damage active fishing gear.

Local, national, and international efforts are needed to address this environmental problem. The Save our Seas Act of 2018 amends and reauthorizes the Marine Debris Act to promote international action, authorize cleanup and response actions, and increase coordination among federal agencies on this topic.

Garbage patches: What and where are they?

Garbage patches are large areas of the ocean where trash, fishing gear, and other marine debris collects. The term "garbage patch" is a misleading nickname, making many believe that garbage patches are "islands of trash" that are visible from afar. These areas are actually made up of debris ranging in size, from microplastics to large bundles of derelict fishing gear.

These patches are formed by large, rotating ocean currents called gyres that pull debris into one location, often to the gyre's center. There are five gyres in the ocean: one in the Indian Ocean, two in the Atlantic Ocean, and two in the Pacific Ocean. Garbage patches of varying sizes are located in each gyre. Due to winds and currents, garbage patches are constantly changing size and shape. The debris making up the garbage patches can be found from the surface of the ocean all the way to the ocean floor.

CAUSES OF MARINE POLLUTION

The inputs and causes of marine pollution can be categorized as the following:

Direct discharge: Pollutants are directly discharged from urban centres or factories into rivers and provided those rivers have outlets into the sea, the hazardous or toxic wastes will find their way into the oceans. Plastics constitute a major component of the hazardous wastes directly discharged into the oceans.As per environmental studies, China, Indonesia, Sri Lanka, Vietnam, Egypt, Thailand, Nigeria are major contributors of ocean plastic pollution.

Mining for minerals such as iron, copper, and zinc is also a major source of marine pollution. Usually, uprooted soil ends up in rivers flowing to the oceans. That itself is not a problem, but minerals that are mined will be a cause for concern as their effects can be poisonous for marine life.

1. **Land run-off:** Surface run-off from farming, industrial, urban activities can carry silt and particles laden with carbon, nitrogen, phosphorus and minerals. Polluted run-off from roads and highways can be a significant source of water pollution in coastal areas.
2. **Ship pollution:** Ship pollution can take many forms. One of the well known varieties are those of oil spills. Oil spills are toxic to marine life and are even more difficult to clean up, taking years to dispose of the sediments that has accumulated.

   While oil spills make the headlines of environmental accidents, discharges from ballast tanks of big oil ships are other sources of pollution from ships. The ballast tanks contain waste waters which can harm marine life.
3. **Atmospheric pollution:** Atmospheric pollution is also the result of marine pollution. Winds blow dust and debris like plastic bags from landfill near ocean bodies.

   Dust from the Sahara, for example, moving around the southern periphery of the subtropical ridge, moves into the Caribbean and Florida during the warm season as the ridge builds and moves northward through the subtropical Atlantic.

Climate change is increasing ocean temperatures as the years go by, leading to carbon dioxide levels rising and altering marine ecosystems and modifying fish distributions. This has impacted communities that depend on fisheries in the process.

4. **Deep Sea Mining:** Deep sea mining is a new mineral retrieval process where the ocean follow is mined for minerals. Precious metals such as silver, gold, copper, cobalt, zinc etc are mined. Like the land variant, deep sea mining releases many pollutants onto the ocean floor, albeit directly.

Since it's a relatively new form of mining, its environmental concerns are yet to be fully ascertained. But what is certain is that deep sea mining is just as harmful as its land counterpart.

Leakages, spills and corrosion will alter the mining areas ecological makeup.

TYPES OF MARINE POLLUTION

Acidification

Oceans are a natural carbon sink, as in it absorbs carbon dioxide from the ocean atmosphere. Since the levels of carbon dioxide are increasing, oceans as a result are becoming more acidic.

Although the long term effects of ocean acidification are yet to be known, it is assumed that corals will be affected, with marine wildlife also suffering in the process

1. **Eutrophication:** Eutrophication is an increase in chemical nutrients, typically compounds containing nitrogen or phosphorus, in an ecosystem. It can result in an increase in the ecosystem's primary productivity (excessive plant growth and decay), and further effects including lack of oxygen and severe reductions in water quality, fish, and other animal populations.

   The biggest culprit are rivers that empty into the ocean, and with it the many chemicals used as fertilizers in agriculture, as well as waste from livestock and humans. An excess of oxygen-depleting chemicals in the water can lead to hypoxia and the creation of a dead zone.

2. **Plastic Debris:** Out of all the debris that float on the surface of oceans, 80% of it is plastic. Since the end of World War II, it has been rapidly accumulating. It is estimated that at least 100,000,000 tonnes may have been accumulated in the ocean ecosystem.

   Plastic debris, when bulky or tangled, is difficult to pass, and may become permanently lodged in the digestive tracts of these animals. Plastic debris also suffocates marine animals fatally. Many marine animals have lost their lives as a result of coming in contact with plastic debris

3. **Toxins:** Apart from plastics, there are particular problems with other toxins that do not disintegrate rapidly in the marine environment. Examples of persistent toxins are PCBs, DDT, TBT, pesticides, furans, dioxins, phenols, and radioactive waste.

Heavy metals are metallic chemical elements that have a relatively high density and are toxic or poisonous at low concentrations. Examples are mercury, lead, nickel, arsenic, and cadmium. Such toxins can accumulate in the tissues of many species of aquatic life in a process called bioaccumulation. They are also known to accumulate in benthic environments, such as estuaries and bay muds: a geological record of human activities of the last century.

WHAT MARINE ENVIRONMENTAL PROBLEMS ARE ASSOCIATED WITH PETROLEUM POLLUTION?

Exploration of oil and gas deposits is now a global industrial activity in the deep ocean. As easily accessible oil and gas resources became depleted, and technology improved, the oil and gas industry expanded into deeper waters in recent decades However, this deep-water expansion has not always been matched by legislation that reflects modern practices of environmental conservation.

There is a clear need to bring together current knowledge of deep-sea ecology, known human impacts on deep-water ecosystems, and the scattered environmental protection measures that exist to date.

Oil destroys the insulating ability of fur-bearing mammals, such as sea otters, and the water repellency of a bird's feathers, thus exposing these creatures to the harsh elements. Without the ability to repel water and insulate from the cold water, birds and mammals will die from hypothermia.

Juvenile sea turtles can also become trapped in oil and mistake it for food. Dolphins and whales can inhale oil, which can affect lungs, immune function and reproduction. Many birds and animals also ingest oil when they try to clean themselves, which can poison them.

Fish, shellfish, and corals may not be exposed immediately, but can come into contact with oil if it is mixed into the water column – shellfish can also be exposed in the intertidal zone. When exposed to oil, adult fish may experience reduced growth, enlarged livers, changes in heart and respiration rates, fin erosion, and reproduction impairment.

Fish eggs and larvae can be especially sensitive to lethal and sublethal impacts. Even when lethal impacts are not observed, oil can make fish and shellfish unsafe for humans to eat.

Petroleum marine fuel spills, which result from damage, transportation accidents and various other industrial and mining activities, are classified as hazardous waste. They are considered to be the most frequent organic pollutants of aquatic ecosystems.

Major Oil Spills of the Maritime World

Almost all the mistakes we make harm the environment and the species on the planet in one way or another. While some human errors affect nature eventually, few make immediate and huge impacts on the environment, with even ending the lives of thousands of species at a time.

When the incidents like wildfire and bombing end up destroying the marine environment in a minute of time on land, it is events such as oil spills that put our seas and oceans into trouble in a short period of time.

Oil spill disasters have been one of the major concerns of the marine world for a long time now. They are commercial and environmental catastrophes.

Previous observations indicated that the number of large spills (>700 tonnes) has decreased significantly over the last 30 years (Table 1).

Table 1. Number of oil spills over 7 tonnes from 1970 to 2013

Year	7-700 tonnes	>700 tonnes	Quantity Spilled x 10^3 tonnes
1970-1974	189	125	1114
1975-1979	342	117	2012
1980-1984	221	41	570
1985-1989	124	48	513
1990-1994	165	48	907
1995-1999	108	25	194
2000-2009	182	-	213
2010-2013	28	-	22

Table 2. Top 22 major oil spills since 1967

Ship name	Year	Location	Spill (10^3) tonnes
Torrey Canyon	1967	Scilly Isles, UK	119
Sea Star	1972	Gulf of Oman	115
Jakob Maersk	1975	Oporto, Portugal	88
Urquiola	1976	La Coruna, Spain	100
Hawaiian Patriot	1977	300 nautical. miles off Honolulu	95
Amoco Cadiz	1978	Off Brittany, France	223
Atlantic Empress	1979	Off Tobago, West Indies	287

Ship name	Year	Location	Spill (10^3) tonnes
Independenta	1979	Bosphorus, Turkey	94
Irenes Serenade	1980	Navarino Bay, Greece	100
Castillo de Bellver	1983	Off Saldanha Bay, South Africa	252
Novo	1985	Off Kharg Island, Iran	70
Odyssey	1988	700 nautical. miles off Nova Scotia, Canada	132
Khark 5	1989	120 nautical. miles off Atlantic coast of Morocco	70
Exxon Valdez	1989	Prince William Sound, Alaska, USA	37
ABT Summer	1991	700 nautical miles off Angola	260
Haven	1991	Genoa, Italy	144
Aegean Sea	1992	La Coruna, Spain	74
Katina P.	1992	Off Maputo, Mozambique	67
Braer	1993	Shetland Islands, UK	85
Sea Empress	1996	Milford Haven, UK	72
Prestige	2002	Off Galicia, Spain	63
Hebel Spirit	2007	South Korea	11

IMPACT OF OIL SPILLS ON PLANKTONIC ORGANISMS

Some zooplankton, such as copepods, euphausiids and mysids, assimilate hydrocarbons directly from seawater and by ingesting oil droplets and oil contaminated food. Zooplankton is a particularly important food resource, especially for baleen whales. It can influence or control the primary productivity by top-down effects in return. Its population dynamic change can influence the biomass of other marine animals like fish by bottom-up effects. The ingestion of oil by these organisms often causes mortality, while surviving organisms often show developmental and reproductive abnormalities.

For the oil WAF, researchers reported that plankton are capable of accumulating PAH due to their great lipophilic abilities. They, therefore, stimulate various harmful effects. The investigators reported that marine plankton is highly sensitive to the petroleum WAF (water accommodated fraction), as the order of median effective/ lethal concentration is as low as lg/L or mg/L. Examinations of the toxicity effect of 10 polycyclic aromatic hydrocarbons associated with the Prestige fuel oil spill on adult copepods (*Oithonadavisae*) revealed that the PAHs had narcotic effects on these organisms.

The impact of petroleum pollution on marine plankton has been a great cause for concern. Researchers summarized the reports regarding the toxic effects of oil water accommodated fraction (WAF) on marine phytoplankton, zooplankton and the early life stages of animals. Generally, oil WAF toxicity enhances with increasing carbonic chain length and benzene ring number.

IMPACT OF OIL SPILLS ON BENTHIC ORGANISMS AND INVERTEBRATES

Benthic invertebrates and higher forms, such as the sand eel and *Ammodvtes americanus* (a main food resource of Atlantic humpback whales), may accumulate petroleum hydrocarbons from water, contaminated sediments and food.

Benthic amphipods are quite sensitive to spilled oil. They are among the first marine animals killed and the slowest to recover. However, most marine crustaceans have a well-developed MFO (mixed function oxygenase) system.

As a result, they are able to metabolize and excrete accumulated hydrocarbons quite rapidly. Bivalve molluscs tend to accumulate petroleum hydrocarbons to higher concentrations and retain them longer than other taxa. This is essentially due to the lack of a mixed function oxygenase system that makes them unable to metabolize the compounds to execrable polar metabolites.

Thus, they are likely to transfer them to their predators. Marine mammals that rely heavily on bivalve molluscs for food, such as the walrus and otter, share a higher risk of ingesting petroleum hydrocarbons. Previous studies have explored the recovery of the invertebrate populations after oil spills. In the intertidal habitat, the biological recovery of the exposed shores is faster than the sheltered shores. This is because strong wave action promotes the removal of contamination and the animals and plants of exposed shores tend to be severely affected. Thus, they are better able to re-colonize an impacted shore quickly.

Invertebrate populations, such as the amphipod sand hopper, *Ampelisca*, are quite sensitive to oil pollution and, for various reasons, are slow to re-populate and 'recover'. For example, the initial impact of the spill in the Bay of Morlaix, following the *Amoco Cadiz* spill in 1978, was to kill off populations of the amphipod sand hopper, *Ampelisca*, which dominated the community. Although the sediment was rapidly purged of the contaminating oil, it has been noted that *Ampelisca*was back to its pre-spill population density after 10 years. The standing crop biomass and productivity was restored much more rapidly as the ecological niches had been occupied by other opportunists like the bivalve *Abra*and the worm *Nephtys*.These had quickly filled the place left by the *Ampelisca*.

Sub-littoral habitats are generally contaminated by sedimentation of oiled particulate material and clean-up for these habitats is not practiced. As a result, the recovery of subtidal communities impacted by oil spills usually takes a longer time. For example, the *Abraalba*bivalve sand community in the Bay of Morlaix, Brittany was severely affected by the *Amoco Cadiz* oil spill (1978). After the spill in 1978, the biomass values for the sand community immediately fell. However, within two years, they recovered to pre-spill levels. Productivity also showed similar trends.

In other studies on oil spill accidents, it has been shown that six months after the Prestige oil spill and clean-up campaign, invertebrate populations of the exposed sandy beaches, notably the isopod *Eurydice*, the spionid polychaete *Scoleleoius squamata*, nemerteans and Diptera, were significantly reduced. Furthermore, their abundance inversely related with the oil pollution gradient. The number of taxa was reduced but not the diversity values. The only clam on these sandy beaches was *Donax*.Before the spill and clean-up, it occurred in six beaches but afterwards, only in one. Upper dry sand communities were particularly reduced due to both oil toxicity and extensive beach grooming that also removed seaweed wrack.

Marine invertebrates are unique living organisms in different aspects. They have been explored as a model for several biological markers as a result of oil spill pollution. Researchers surveyed the marine invertebrate mussels *(Mytilus edulis)* exposed in situ to the oil that came ashore after the wreck of the '*Erika*' tanker on the Brittany (France) coast in December, 1999. The mussel response was assessed using a set of biomarkers (acetylcholinesterase (AChE), glutathione S-transferase (GST), catalase (CAT), malondialdehyde (MDA) and deoxy-ribonucleic acid (DNA) adducts, related to the metabolism of the organic contaminants. The results of a series of validation tests revealed that there were no significant reductions in the GST or CAT levels. Six months immediately following the accident, observations indicated that the DNA adducts and MDA levels were high and the levels of AChE were significantly lower during the first year of the survey. This suggested a general stress. A simple multivariate graphic method - the integrated biomarker response index - was used to combine four of the five validated biomarkers and quantify the degree of impact on mussels at different sites. The results indicated that mussel populations were affected by the oil spill during only the first year after the accident.

IMPACT OF OIL SPILLS ON CORAL REEFS

In addition, recreational attractions for divers, coral reefs are considered to be important constituents of marine ecosystems. This is because they are important nurseries for shrimp, fish and other animals. The aquatic organisms that live within and around the coral reefs are at risk of exposure to the toxic substances within oil, as well as smothering. They are rapidly deteriorating because of a variety of environmental and anthropogenic pressures. Thus, they are suffering significant changes in diversity, species abundance and habitat structure worldwide.

Oil dispersants are potentially harmful to marine life including coral reefs. In a study using coral nubbins in coral reef ecotoxicology testing, found that dispersed oil and oil dispersants are harmful to soft and hard coral species at early life stages. The investigators also employed a 'nubbin assay' on more than 10, 000 coral fragments to evaluate the short- and long-term impacts of dispersed oil fractions (DOFs) from six commercial dispersants (Slickgone, Petrotech, Inipol, Biorieco, Emulgal and Dispolen) and the dispersants and water-soluble-fractions (WSFs) of Egyptian crude oil on two Indo-Pacific branching coral species, *Stylophora pistillata* and *Pocillopora damicornis*. They found that the dispersant concentrations recommended by the manufacturer were highly toxic and resulted in mortality of all nubbins.

The dispersed oil and the dispersants were significantly more toxic than the crude oil WSFs. As corals are very sensitive to oil detergents and dispersed oil, the results of these assays indicated the negative and harmful effect of using any oil dispersant in coral reefs and in the area closely around them. These dispersants were rated based on their eco-toxicological impacts on the corals. Observations indicated a scale from the least to the most harmful agent, as follows: Slickgone > Petrotech > Inipol > Biorieco > Emulgal > Dispolen.

IMPACT OF OIL SPILLS ON FISH

Due to the well-developed hepatic mixed function oxygenase (MFO) system, in addition to the reactivity of the metabolites that would not be released in a toxic form during digestion and absorption, most fish, even in heavily oil-contaminated environments, do not accumulate and retain high concentrations of petroleum hydrocarbons. Thus, they are not likely to transfer them to predators. Thus, no serious threat is predicted.

Generally, marine carnivores are inefficient assimilators of petroleum compounds in food. For this reason, and because all prey species are able to release hydrocarbons from their tissues, the marine food chain biomagnification does not occur. Thus, there is an indirect correlation between a marine mammal's trophic level and the concentration of residues that it might consume. In fact, as top carnivores that feed on large pelagic fish and seals, polar bears and killer whales are less likely to be exposed to petroleum in their food than other species, such as walrus and baleen whales, which feed on zooplankton and benthic invertebrates.

Experiments that explain the effect of oil spills on the early stages of fish were demonstrated by [56]. They found that chronic exposure of juvenile pink salmon (*Oncorhynchus gorbuscha*) to Alaska North slope crude oil resulted in a variety of responses. These included melanosis, erratic swimming, loss of equilibrium, reduced mobility and startled response.

Interestingly, the effects were not enhanced by the phototoxicity from UV irradiation, presumably of the highly pigmented nature of the fish's skin (in contrast to phototoxicity in translucent early life stages of marine invertebrates). When they examined the phototoxicity of the water accommodated fractions of weathered crude oil, they found that pink salmon may be at less risk of photoenhanced toxicity, compared to the more translucent early-life stages of several other Alaskan species.

IMPACT OF OIL SPILLS ON SEABIRDS

As one of the major routes of exposure, physical contact usually affects birds. For example, thousands of African penguins (*Spheniscus demerus*) were oiled following the 2000 Treasure oil spill in South Africa.

An evaluation of the impact of oil spills on seabirds has not been fully appreciated during incidents, despite pressure from the public concern, media and other interested parties for precise and up to date information on the damage. Consequently, the approximate numbers of seabird casualties involved in many major spills have only been estimated, while impacts at the population level have been difficult to determine. Natural variation and the huge range of factors that influence bird population statistics make it difficult to assess the impact of oil spill on sea birds.

Researchers, in their efforts to assess the impact of oil spills on seabirds in Europe and in North America, the investigators reported that there are two inter-linked aspects to dealing with oiled seabirds during major spills. The first is a relatively expensive and logistically complicated process.

This involves birds found alive and the humanitarian efforts made to clean, rehabilitate and successfully release them into the wild. The second involves assessing the likely impact of the spill on the populations of those species affected.

The relation between the size of an oil spill and the number of seabird casualties is not directly proportional. Moreover, estimates of the number of seabird deaths from oil slicks are highly speculative. This is because an unknown number of oiled birds may die at sea and not reach the coast. For example, following the *Exxon Valdez*oil spill, over 35,000 seabird carcasses were recovered in the northern Gulf of Alaska.

Researchers argued that the rapid recovery of the murre breeding colonies in Alaska and the number of dead birds might be lower than was estimated. They also suggested that breeding pairs may have been replaced by younger birds that came in from the open sea. On the other hand, researchers argued that the impacts of the spill may have been diffused over a large area, permitting local recovery and making it difficult to detect any changes in local abundance or habitat occupancy. Nine years after the Exxon Valdez oil spill, in their long-term effects observation, researchers claimed that the populations of most bird species have not recovered and others still show potential population effects. However, the report is at variance with other findings.

During the winter season of 2003, approximately 64,000 tonnes of Prestige heavy fuel oil spilled off Galacia, Spain. As a result, 10 % of the European shag (cormorants, *Phalacrocorax aristotellid*) were killed. In turn, this resulted in a 50% reduction in the 2003 breeding success of oiled colonies, compared to the unoiled colonies. Researchers found that the kidneys of 32 oiled shag had extensive tissue damage and haemorrhaging. They further found that the oiled shag were heavily infected with an eimeriorin Apicomplex coccidian, which under normal circumstances is not pathogenic.

IMPACT OF OIL SPILLS ON MARINE MAMMALS

Marine mammals include bottlenose dolphins, fins, humpbacks, rights, sei whales, sperm whales, manatees, cetaceans, seals, sea otters and pinnipeds. As previously indicated, the physical contact of oil with furred mammals usually affects these animals because they rely on their outer coats for buoyancy and warmth. Consequently, these animals often succumb to hypothermia, drowning and smothering when oil flattens and adheres to the outer layer.

As part of their activities, all marine mammals spend a considerable amount of time at the surface. Here, they swim, breathe, feed or rest. Thus, the possibility of their contact with a surface slick, water-in-oil emulsion, or tar balls, is high. In heavy pelage marine mammals, such as fur seals, sea otters and polar bears, this contact may lead to fouling.

Polar bears and otters groom themselves regularly as a means of maintaining the insulating properties of the fur and may, thereby, ingest oil. Animals with smooth surfaces or relatively little to no pelage, such as whales, dolphins, manatees and most seals, have an advantage as oil would have fewer tendencies to adhere to their surface.

Some baleen whales are skim-feeders, i.e., they eat at the surface. When in an area of slick or tar balls, this behaviour can lead to foul in the feeding apparatus. Tarry residues, in particular, can coat the baleen plates.

Animals, such as narwhals, belugas, ringed seals, walruses and polar bears in Polar Regions, spend most of their time at the ice edge in leads, polynyas and breathing holes. This is where spilled oil tends to accumulate.

Oil that contaminates a shore is likely to severely affect pinnipeds. Pinnipeds require such areas for nursery and, to a lesser extent, otters and bears. Some of the oil is eventually returned in subtidal sediments, where it may transfer to gray whales, walrus and some seals. Such species feed heavily on benthic animals.

When marine mammals encounter fresh oil, they are likely to inhale volatile hydrocarbons evaporating from the surface slick. These volatile fractions contain toxic monoaromatic hydrocarbons (benzene, toluene and xylenes) and low molecular weight aliphatics with anaesthetic properties. The inhalation of these volatile hydrocarbon compounds is potentially harmful.

The inhalation of concentrated petroleum vapours can cause the inflammation of and damage to the mucus membranes of airways, lung congestion or even pneumonia. Volatile benzene and toluene, which can be inhaled, can be transferred rapidly from the bloodstream into the lungs. Furthermore, they can accumulate from the blood into the brain and liver, causing neurological disorders and liver damage.

Marine mammals are probably poor accumulators to oil directly from the solution or dispersion in the water column. This is because the skin of cetaceans is relatively impermeable to oil. Additionally, most marine mammals do not drink large volumes of seawater. Thus, a significant accumulation of hydrocarbons by this route is unlikely to occur.

Researcher, in his study of 'Physiologic and Toxicologic Effects on Pinnipeds', indicated that pinnipeds are inappropriately sensitive to the harmful properties of oil. Incidental ingestion during feeding, exposure to vapour concentrations and surface contamination with relatively fresh oil does not appear to cause a disaster.

However, Pinnipeds trapped near the source of a spill or those which are forced to emerge in heavy accumulations of oil in leads and around rookeries exhibit the most severe effects. Experimental studies by the same investigators on fur seals indicated that surface fouling decreases the insulation value of the pelt. This can potentially lead to thermal and energetic stress. Furthermore, the sensitivity to the effects of oil exposure may be high in species and groups that are compromised by pre-existing disease, or stressed by pressures of an unfavourable habitat, intra-specific competition or unusual environmental conditions.

IMPACT OF OIL SPILLS ON MARINE PLANTS

In several aspects, aquatic plants are important to the functioning of ecosystems. These include the fact that they are oxygen producers, their ability to sequester carbon and for their base position in aquatic food chains. In addition, they serve as nursery, feeding and breeding habitats for a variety of animal and plant species, including recreationally and commercially important fish.

Plants and animals are affected by the oil in which they come into contact with as a result of an oil spill. In their review of toxicities of oils, dispersants and dispersed oils to algae and aquatic plants, researchers summarized the reported phytotoxicities of oils, dispersants and their combinations to aquatic plants. They assessed the ability of the reviewed database to support toxicity predictions and evidence-based risk assessments.

The phytotoxicity database mainly includes the results of research conducted after oil spills to marine waters. The toxicity of at least 41 crude oils and 56 dispersants were recorded. At least 107 response parameters were monitored for 85 species of unicellular and multicellular algae, 28 wetland plants, 13 mangroves and nine seagrasses. Due to experimental diversity, the effect concentrations available from this toxicity database are varied and diverse. As a result, there are restricted phytotoxicity predictions and identification of sensitive species, life stages and response parameters. Thus, the impact of toxicity of petrochemicals and dispersants on aquatic plants was not supported by this database.

Researchers recommended a proactive and experimentally-consistent approach to provide the threshold toxic effect concentrations for sensitive life stages of aquatic plants inhabiting diverse ecosystems.

The effects of heavy fuel oil contamination on the growth and the development of *Salicorniafragilis*Ball and Tutin, a salt-marsh edible species, were studied as greenhouse experiments by researchers. Phytotoxicity assessments and PAH shoots assays were followed to measure the impact of petroleum on plant development. Syptoms like chlorosis, yellowing, growth reduction and perturbations in developmental parameters were visually observed. In this study, shoot coating appeared to be less than through soil and the investigators observed more marked effects on plants as an indication of the degree of pollution.

However, a significant bioaccumulation of PAHs in shoot tissues was also found, even at very low levels of contamination. These highly related to the conditions of exposure to oil. This strong relationship between the PAH contents of *Salicornia*plants and growth reduction suggest a chemical toxicity of fuel oil. The investigators concluded that the type and degree of fuel oil contamination are two important factors in controlling the impact of fuel oil on *S. fragilis.*

In 1986, more than 8 million litres of crude oil spilled into a complex region of mangroves, seagrasses and coral reefs, just east of the Caribbean entrance to the Panama Canal. Intertidal mangroves, sea grasses, algae and associated invertebrates were covered by oil and died soon after. Investigators reported that seedlings of red mangrove, *Rhizophoram angle*, which were transplanted to heavily oiled sites, did not produce new leaves. This contrasted the transplants at an unoiled site. Entire beds of intertidal seagrass *Thalassia testudinum* were killed on some heavily oiled reef flats, as shown by the abundant oil-covered dead leaves washed ashore, as well as the dead, but intact, root-rhizome mats. In contrast, the subtidal *Thalassia* survived everywhere after the spill. Having said this, in the heavily oiled areas, the leaves of the subtidal *Thalassia* became brown and heavily fouled by algae for several months.

Another example of the oil spill effect on mangrove plants is December, 2000. In this case, 500 mangrove saplings in a 6.34 ha reserve in Hong Kong were subjected to a smuggled fuel oil spill. More than 80% died in root-zone sediments containing 60 - 80 ug/g total petroleum hydrocarbons (TPH). However, one year later (December, 2001), the injured survivors had recovered and re-grown, with root-zone sediment TPH concentrations approaching urban 'background' values of 40 μg/g TPH.

IMPACT OF OIL SPILLS ON CYANOBACTERIA AND OTHER MICRO-ORGANISMS

There are numerous factors that determine the microbial response to marine oil spills. These include the oil composition and degree of weathering, as well as the environmental conditions, particularly temperature and nutrient concentrations. Reference, for example, reviewed the different factors affecting the biodegradation of the petroleum hydrocarbons by microorganisms and how environmental and biological factors could determine both the rate at which and extent to which hydrocarbons are removed from the environment.

When crude oil is introduced into seawater, the microbial community changes and consists of multiple co-existing species. These can be explained by resource sharing. Researchers, in their review paper, reported that the diverse array of hydrocarbons present in crude oil requires resource partitioning by microbial populations, as well as microbial modification of oil components and the surrounding environment, will lead to temporal succession. The reviewers observed a network of direct and indirect interactions within and between species, even when just one type of hydrocarbon is present. They also provided a schematic illustration of some of the interactions involved in hydrocarbon biodegradation. Elements of these interactions were stated in several studies reviewed by them.

CHEMICAL POLLUTANTS IN THE MARINE ENVIRONMENT

Chemical pollution is when fluids, often toxic, get released into the water. Chemicals can find their way into the sea in many ways and come from all sorts of places.

Industry - Sometimes factories and industry will allow waste products to flow into streams and rivers which eventually get to the sea. Sometimes chemicals are dumped at sea. Chemicals are used in millions of processes whether it is making food, plastic toys, CDs building materials, gadgets or electricals. Through rivers and streams some pollutants enter marine food chains, building up their concentrations until they reach toxic levels.

It often takes human casualties to alert us to pollution and such was the case in Minimata Bay in Japan when many people died as a result of a pollutant building up in food chains. A factory was discharging waste containing methyl mercury in low concentrations into the sea and as this pollutant passed through food chains it became more concentrated in the tissues of marine organisms until it reached toxic levels. As a consequence over a thousand people died from eating fish and shellfish contaminated with mercury and around two thousand people suffered from mercury poisoning.

Farming - Pesticides and herbicides are chemicals sprayed onto food crops to control insects and weeds. These chemicals are often toxic as they are designed to kill or repel animals and contain some synthetic (man-made) chemicals. When sprayed onto crops these substances get left in the soil and can easily find their way into freshwater streams and rivers and end up in the sea. Once in the sea these chemicals often do not disappear by biodegrading as they are resistant to natural breakdown processes but stay in the sea and enter the food chain.

Medicine - Sometimes medicines that humans and animals take also contain bad chemicals and these make their way into the sea through our toilets, into sewage treatment plants and then out into rivers into the sea. For us these chemicals are tolerable in small doses but in aquatic environments they can build up and cause fertility problems in fish.

At home - In our homes we use many different sorts of cleaning, washing and polishing products which all contain various chemicals which often end up down the drain when washed out of dishclothes or used in the bath or kitchen sink. These chemicals may be harmful to wildlife and marine environments as they can contain harmful substances such as sodium hypochlorite, petroleum distillates, phenol and cresol, ammonia and formaldehyde. Buying cleaning products that contain natural soluble/biodegradable ingredients helps to prevent these nasty fluids from ending up in natural water sources.

NON-POINT SOURCE (NPS) POLLUTION

NPS pollution generally results from land runoff, precipitation, atmospheric deposition, drainage, seepage or hydrologic modification. NPS pollution, unlike pollution from industrial and sewage treatment plants, comes from many diffuse sources. NPS pollution is caused by rainfall or snowmelt moving over and through the ground. As the runoff moves, it picks up and carries away natural and human-made pollutants, finally depositing them into lakes, rivers, wetlands, coastal waters and ground waters.

Nonpoint source pollution can include:

- Excess fertilizers, herbicides and insecticides from agricultural lands and residential areas
- Oil, grease and toxic chemicals from urban runoff and energy production
- Sediment from improperly managed construction sites, crop and forest lands, and eroding streambanks

- Salt from irrigation practices and acid drainage from abandoned mines
- Bacteria and nutrients from livestock, pet wastes and faulty septic systems
- Atmospheric deposition and hydromodification

States report that nonpoint source pollution is the leading remaining cause of water quality problems. The effects of nonpoint source pollutants on specific waters vary and may not always be fully assessed. However, we know that these pollutants have harmful effects on drinking water supplies, recreation, fisheries and wildlife.

Nonpoint Sources vs. Point Sources

The term "nonpoint source" is defined to mean any source of water pollution that does not meet the legal definition of "point source" in section 502(14) of the Clean Water Act:

> The term "point source" means any discernible, confined and discrete conveyance, including but not limited to any pipe, ditch, channel, tunnel, conduit, well, discrete fissure, container, rolling stock, concentrated animal feeding operation, or vessel or other floating craft, from which pollutants are or may be discharged. This term does not include agricultural storm water discharges and return flows from irrigated agriculture.

What You Can Do to Prevent Nonpoint Source (NPS) Pollution

In Urban Environments

- Keep litter, pet wastes, leaves and debris out of street gutters and storm drains—these outlets drain directly to lake, streams, rivers and wetlands.
- Apply lawn and garden chemicals sparingly and according to directions.
- Dispose of used oil, antifreeze, paints and other household chemicals properly—not in storm sewers or drains. If your community does not already have a program for collecting household hazardous wastes, ask your local government to establish one.
- Clean up spilled brake fluid, oil, grease and antifreeze. Do not hose them into the street where they can eventually reach local streams and lakes.
- Control soil erosion on your property by planting ground cover and stabilizing erosion-prone areas.
- Encourage local government officials to develop construction erosion and sediment control ordinances in your community.
- Have your septic system inspected and pumped, at a minimum every three to five years, so that it operates properly.

- Purchase household detergents and cleaners that are low in phosphorous to reduce the amount of nutrients discharged into our lakes, streams and coastal waters.

Mining

- Become involved in local mining issues by voicing your concerns about acid mine drainage and reclamation projects in your area.

Forestry

- Use proper logging and erosion control practices on your forest lands by ensuring proper construction, maintenance, and closure of logging roads and skid trails.
- Report questionable logging practices to state and federal forestry and state water quality agencies.

Agriculture

Manage animal manures to minimize losses to surface water and ground water.

- Reduce soil erosion and nturient loss by using appropriate conservation practice systems and other applicable best management practices.
- Use planned grazing systems on pasture and rangeland.
- Dispose of pesticides, containers, and tank rinsate in an approved manner.
- Work with conservation partners locally including Soil and Water Conservation Districts to understand local strategies.

THE PROBLEMS WITH OCEAN DUMPING

Although policies on ocean dumping in the recent past took an "out of sight- out of mind" approach, it is now known that accumulation of waste in the ocean is detrimental to marine and human health. Another unwanted effect is eutrophication. A biological process where dissolved nutrients cause oxygen-depleting bacteria and plants to proliferate creating a hypoxic, or oxygen poor, environment that kills marine life. In addition to eutrophication, ocean dumping can destroy entire habitats and ecosystems when excess sediment builds up and toxins are released. Although ocean dumping is now managed to some degree, and dumping in critical habitats and at critical times is regulated, toxins are still spread by ocean currents. Alternatives to ocean dumping include recycling, producing less wasteful products, saving energy and changing the dangerous material into more benign waste.

According to the United Nations Group of Experts on the Scientific Aspects of Marine Pollution, the amount of ocean dumping actually brings in less pollution than maritime transportation, atmospheric pollution, and land based pollution like run-off. However, when waste is dumped it is often close to the coast and very concentrated.

Waste dumped into the ocean is categorized into the black list, the gray list, and the white list. On the black list are organohalogen compounds, mercury compounds and pure mercury, cadmium compounds and pure cadmium, any type of plastic, crude oil and oil products, refined petroleum and residue, highly radioactive waste, any material made for biological or chemical warfare.

The gray list includes water highly contaminated with arsenic, copper, lead, zinc, organosilicon compounds, any type of cyanide, flouride, pesticides, pesticide by-products, acids and bases, beryllium, chromium, nickel and nickel compounds, vanadium, scrap metal, containers, bulky wastes, lower level radioactive material and any material that will affect the ecosystem due to the amount in which it is dumped.

The white list includes all other materials not mentioned on the other two lists. The white list was developed to ensure that materials on this list are safe and will not be dumped on vulnerable areas such as coral reefs.

Issues associated with Ocean Dumping

Incineration has been used to control dangerous chemical waste being dumped into the Ocean. This practice began in 1969 and became very popular by the late 1970's. By the mid-1980's, about 100,000 tons of waste was incinerated before it was dumped. This process has been studied to determine whether the practice is safe and effective. When dangerous waste is burned, it can produce smoke full of hazardous chemicals and may possibly spill into the ocean. After thorough evaluation, it was determined in 1989 that incineration is not a viable method of reducing the amount of waste dumped into the ocean because of the smoke released, and therefore, almost all the waste burning vessels were grounded. The Protocol of 1996 banned burning waste at sea altogether.

The majority of nuclear waste in the ocean comes from six submarine reactors, one nuclear icebreaker reactor, and damaged nuclear fuel in the Kara Sea. The rest of the nuclear material in the ocean is solid nuclear waste in concrete drums.

Although some claim the risk to human health is small, the long-term affects of nuclear dumping are not known, and some estimate up to 1000 deaths in the next 10,000 years as a result of evaporated nuclear waste.

In 1995, a Global Waste Survey and the National Waste Management Profiles inventoried waste dumped worldwide to determine what countries were dumping waste and how much was going into the ocean. Countries that exceeded an acceptable level would then be assisted in the development of a workable plan to dispose of their waste.

The impact of a global ban on ocean dumping of industrial waste was determined in the Global Waste Survey Final Report the same year. In addition to giving the impact for every nation, the report also concluded that the unregulated disposal of waste, pollution of water, and buildup of materials in the ocean were serious problems for a multitude of countries.

The report also concluded that dumping industrial waste anywhere in the ocean is like dumping it anywhere on land. The dumping of industrial waste had reached unacceptable levels in some regions, particularly in developing countries that lacked the resources to dispose of their waste properly.

Enforcement of regulations was also a problem in these areas, so it was necessary to assist the countries in the implementation of a waste disposal strategy.

Areas where high levels of illegal dumping occur are commonly areas that have no way of implementing a better strategy. A global ban on ocean dumping is not enough to eliminate the practice, and it will require all governments to have a workable plan to reduce the amount of waste, recycle some waste, and learn how to modify waste that is dumped into the ocean so it is less harmful.

The Effects of Runoff & Pollution

The ocean is the basin that catches almost all the water in the world. Eventually, water evaporates from the ocean, leaves the salt behind, and becomes rainfall over land. Water from melted snow ends up in rivers, which flows through estuaries and meets up with saltwater.

Fertilizers, pesticides, and oil, mostly from farms, seep into the ground after a rain and then stream into rivers and ultimately into the ocean. Feedlots in the United States exceed the amount of human waste with more than 500 millions tons of manure each year.

Not only does the waste flow into the ocean, but it also encourages algal blooms to clog up the waterways, causing meadows of seagrass, kelp beds and entire ecosystems to die. A zone without any life remaining is referred to as a dead zone and can be the size of entire states, like in coastal zones of Texas and Louisiana.

All major bays and estuaries now contain dead zones from pollution run-off and in 2008, 405 dead zones were counted worldwide. Pollutants like mercury, DDT, PCBs and pesticides are becoming more common in seafood meant for the dinner table and cause birth defects, cancer and neurological problems—especially in infants.

Recreation

Water recreation is another aspect of human life compromised by marine pollution from human activities like roads, shopping areas, and development in general.

Swimming is becoming unsafe, in 2008, of the 3,740 coastal beaches that were monitored, 1210 had at least one advisory or closure due to contamination from pollutants. Developed areas like parking lots enable runoff to occur at a much higher volume than a naturally absorbent field. Even simply driving a car or making a house warm can leak 28 million gallons of oil into lakes, streams and rivers.

Environmental Laws

In addition to the lack of underwater national parks, there is no universal law like the Clean Air Act or the Clean Water Act to protect the United States ocean territory. Instead, there are many different laws like the Magnuson-Stevens Fishery Conservation and Management Act, which only apply to certain aspects of overfishing and are relatively ineffective. The act developed in the 1970's is not based on scientific findings and is regulated instead by the regional fisheries council. In 2000, the Oceans Act was implemented as a way to create a policy similar to the nationwide laws protecting natural resources on land. However, this act still needs further development and, like many of the conservation laws that exist at this time, it needs to be enforced.

Development

A major factor in the destruction of habitats is the development of coasts and the loss of water quality.

Usually, coastal marshes can help control pollution by filtering out normal pollutants and providing a safe haven for young fish and other wildlife. However, these marshes are vanishing at 20,000 acres a year and most of the seagrass cover is being lost very quickly in places like Tampa Bay and the Mississippi Sound.

Another crucial ecosystem is the coral reef, a place of great biodiversity, recreational value, a sanctuary for most young fish and very important to the protection of coastlines during storms. Unlike terrestrial national parks that are in full view, pristine underwater areas are rarely seen by the United States government as high priority.

THE OCEAN ECOSYSTEM

The ocean is a complex and interwoven ecosystem with each biotic and abiotic factor influencing every other component directly or indirectly. When one habitat vanishes, organisms that rely on that niche can no longer survive.

Animals like sea turtles, manatees, fish, shrimps and crabs rely on seagrass for survival. Unfortunately, seagrass is sensitive to dredging of channels, pollution and development and is being lost at an alarming rate. Since the ocean is an interlaced ecosystem, the loss of one species like sea otters can cause sea urchins (their prey) to soar in numbers. The loss of biodiversity in the oceans is more critical than simply losing a source of wonder for future generations. Loss of biodiversity has the power to affect human survival in the future in a profound way. For example, species like sea sponges have been found to hold chemicals capable of beating cancer and viruses, but they will be lost if pollution from runoff and ocean dumping is not curbed dramatically.

HOW CAN YOU HELP OUR OCEAN?

Ocean pollution is a raging problem that needs to be solved as early as possible. With the world's oceans getting polluted, the marine ecosystems are getting severely disturbed. Not only that, but the world's water reserves are also becoming more and more limited. In these situations, it is essential that we find ways of solving ocean pollution. Some of the ways are as follows:

Reducing the Use of Plastic Products

Believe it or not, plastic wastes form the largest portion of ocean pollutants. Out of the 260 million tons of plastic produced each year globally, approximately 10% ends up in the oceans.

These plastics then take thousands of years to decompose, during which time it posses a terrible threat to life in the oceans. Thus, reduction in the use of plastic products could help in significantly reducing the rates of ocean pollution.

Use Reusable Bottles and Cutlery

Use and throw bottles and cutlery, too, are an immense contributor to ocean pollution. We must not forget that most of the non-reusable bottles and cutlery are made out of either plastic or Styrofoam. Both these materials require a few hundred or thousand years to decompose.

A plastic bottle would require about 450 years to decompose completely. In the meanwhile, it would remain inside the ocean, releasing toxins that would poison the marine life.

Recycle Whatever You Can

The 3 R's are definitely the savior of ocean health. With the increase in population, it is true that the waste produced would also increase significantly. As finding dumping grounds become more and more difficult, a lot of the wastes find their way to the ocean beds. These wastes that are dumped in the ocean remain there for a long time, affecting marine life.

One of the useful ways to reduce waste production is recycling. Before throwing things away, we could check if something is recyclable. Then the products that can be recycled can be taken to the recycling center nearby.

Stop Littering the Beach, and Start Cleaning It

The beach is undoubtedly one of the loveliest places to visit and hang out in. As a result of this, there is a lot of littering around as well. The first and foremost thing that really needs to stop is this. Abundant dustbins should be placed on the beaches.

If someone is seen littering, they must be stopped immediately, and certain strict rules and regulations must be followed at all times. Also, if the beach seems untidy, we could pick up the wastes and throw it in the dustbin.

Reducing the Use of Chemical Fertilizers

Runoffs, too, pollute the oceans immensely. To prevent this, the use of chemical fertilizers must also be controlled and regulated. It must be remembered that excess use of chemical fertilizers harms not only the soil but also the water bodies nearby and, ultimately, the ocean. These runoffs are so toxic that they can very easily kill marine life.

Reducing the Energy Use

As the population increases, our demand for energy increases too. A major chunk of this energy is produced from petroleum. This petroleum is obtained by the drilling of the ocean beds. Any spillage during the procedure could harm marine life. Not only that, but the procedure itself also violates the marine ecological balance. So by controlling our energy use, we could also solve the problem of ocean pollution.

CHAPTER

MARINE LIFE AND THE MARINE ENVIRONMENT 12

Marine waters cover two-thirds of the surface of the Earth. In some places the ocean is deeper than Mount Everest is high; for example, the Mariana Trench and the Tonga Trench in the western part of the Pacific Ocean reach depths in excess of 10,000 metres (32,800 feet). Within this ocean habitat live a wide variety of organisms that have evolved in response to various features of their environs.

ORIGINS OF MARINE LIFE

The Earth formed approximately 4.5 billion years ago. As it cooled, water in the atmosphere condensed and the Earth was pummeled with torrential rains, which filled its great basins, forming seas. The primeval atmosphere and waters harboured the inorganic components hydrogen, methane, ammonia, and water. These substances are thought to have combined to form the first organic compounds when sparked by electrical discharges of lightning. Some of the earliest known organisms are cyanobacteria (formerly referred to as blue-green algae). Evidence of these early photosynthetic prokaryotes has been found in Australia in Precambrian marine sediments called stromatolites that are approximately 3 billion years old.

Although the diversity of life-forms observed in modern oceans did not appear until much later, during the Precambrian (about 4.6 billion to 542 million years ago) many kinds of bacteria, algae, protozoa, and primitive metazoa evolved to exploit the early marine habitats of the world. During the Cambrian Period (about 542 million to 488 million years ago) a major radiation of life occurred in the oceans. Fossils of familiar organisms such as cnidaria (e.g., jellyfish), echinoderms (e.g., feather stars), precursors of the fishes (e.g., the protochordate *Pikaia* from the Burgess Shale of Canada), and other vertebrates are found in marine sediments of this age. The first fossil fishes are found in sediments from the Ordovician Period (about 488 million to 444 million years ago). Changes in the physical conditions of the ocean that are thought to have occurred in the Precambrian—an increase in the concentration of oxygen in seawater and a buildup of the ozone layer that reduced dangerous ultraviolet radiation—may have facilitated the increase and dispersal of living things.

Marine life refers to all the animals, plants and organisms that live in Earth's saltwater seas and oceans. From the smallest plankton to the largest whale, all organisms play a role in the healthy function of these amazing, complex ecosystems.

Coral Reef Ecosystems

Coral reefs are diverse, colorful and home to millions of marine species across the planet. They provide a natural barrier to the power of the ocean, protecting coastal communities from extreme weather events such as tsunamis. As food chains, they are extremely important, providing an area in which marine species can live, feed, raise their young and thrive.

Unfortunately, they are also highly vulnerable to ocean temperature change, a direct result of climate breakdown. In 2016, Australia's Great Barrier Reef experienced a 30 percent loss of coral reef due to a nine-month marine heatwave, the largest reef loss ever recorded. The fishing of reefs has also led to species loss, as food chains are interrupted and predators cannot survive.

Some examples of the largest reef ecosystems on earth include:

- The Great Barrier Reef, Australia
- The Red Sea Reef, Egypt, Sudan and Eritrea
- New Caledonia Barrier Reef, southwest Pacific Ocean
- Florida Reef, USA
- Andros Coral Reef, the Bahamas
- Ocean Fishes

There are thousands of different species of ocean fish, yet populations are continually declining as we allow commercial fleets to fish our oceans. Atlantic cod, for example, can live up to 25 years, yet their populations are now at critically low levels. Some other ocean fish include:

- Tunas
- Mackerels
- Sardines
- Mahi Mahis
- Barramundis
- Anchovies

Marine Mammals

Marine mammals are generally larger sea creatures, who live underwater but need to breathe air also. Around the world these marine mammals still lose their lives unnecessarily due to habitat loss, being caught and killed by fishing fleets ("fishing bycatch"), hunting and noise pollution.

- Marine mammals include:
- Dolphins
- Whales
- Seals
- Otters
- Porpoises
- Manatees

Sea Turtles And Reptiles

Reptile life in the sea is largely made up of turtle species such as:

- Loggerhead turtles
- Hawksbill sea turtles
- Green sea turtles
- Leatherback sea turtles

All seven species of sea turtles are endangered or critically endangered with The World Wildlife Fund stating that "the single biggest threat to most sea turtles is fishing gear."

Marine iguanas and saltwater crocodiles are also ocean-dwelling species of reptile. Reptilian life in the sea is particularly vulnerable to pollutants such as plastic, as these animals live in shallow seas and come into contact with human populations often.

Cephalopods, Crustaceans And Shellfish

Cephalopods, crustaceans and shellfish play an important role in ocean ecosystems and the marine food chains, and includes:

- Octopi
- Squids
- Lobsters
- Crabs
- Cuttlefishes
- Shrimps
- Prawns

They are particularly vulnerable to increased ocean acidification, a direct result of increased human-generated greenhouse gas emissions.

Seabirds

Seabirds are found on coastlines, in coastal waters and far out at sea, and include species such as:

- Penguins
- Albatrosses
- Fish eagles
- Puffins

Importance of Marine Life

The oceans provide 50 percent of the world's oxygen and they provide essential ecosystem services that allow our planet to function in a healthy way. For example, ocean currents govern our weather systems, and if they were to be interrupted, the extreme weather consequences would be drastic. The oceans can only provide these services by maintaining delicately balanced ecosystems, made up of food chains and biological systems. Human activity risks the shutdown of essential ecosystem services that we rely on to survive.

One example is coral reefs. When we fish around coral reefs, those reefs' delicate ecosystems are damaged and often die. We then lose the essential service a coral reef provides such as protection from erosion and weather, biochemical regulation, and recreation. This is just one of the reasons why taking care of the oceans is essential for our own wellbeing.

Types of Marine Ecosystems

Scientists divide marine ecosystems into several broad categories, although there are discrepancies depending on the source about what qualifies as a marine ecosystem. The number of marine ecosystems is actively debated. Although there is some disagreement, several types of marine ecosystems are largely agreed on: estuaries, salt marshes, mangrove forests, coral reefs, the open ocean, and the deep-sea ocean.

An estuary is a coastal zone where oceans meet rivers. Here, nutrients and salts from the ocean mix with those from the river in regions sheltered from extreme weather. As a result, estuaries are among the most productive places on Earth and support many types of life. In addition, because they are located where rivers join the ocean, estuaries have traditionally supported many human communities and activities like fishing, shipping, and transportation.

While estuaries form where ocean meets rivers, salt marshes occur where oceans meet land. These places are rich in nutrients from sediment brought in by the ocean.

Marshes are regularly flooded by high tides, making the surrounding ground wet and salty. As a result, the soil is low in oxygen and filled with decomposing matter. These ecosystems are dominated by low-growing shrubs and grasses.

Another coastal ecosystem is the mangrove forest. Mangrove forests are found in tropical areas. These ecosystems frequently flood with ocean water, submerging the roots of mangrove trees. The root systems of mangroves filter out salt and sit above ground to access oxygen. These trees provide a home for a variety of species. Animals, such as fish, crabs, shrimp, reptiles, and amphibians, live among the mangrove's roots while its canopy provides a nesting site for birds.

A bit farther out into the tropical sea are coral reefs, euphotic-zone ecosystems built from the exoskeleton secreted by coral polyps. These exoskeletons form complex structures that shelter many different organisms. Coral reefs are extremely diverse ecosystems that host sponges, crustaceans, mollusks, fish, turtles, sharks, dolphins, and many more creatures. By some counts, coral reefs can account for a quarter of all ocean species.

Beyond the coral reefs lies the open ocean. Open ocean ecosystems vary widely as the depth of the ocean changes. At the surface of the ocean, the euphotic zone, the ecosystem receives plenty of light and oxygen, is fairly warm, and supports many photosynthetic organisms. Many of the organisms that we associate with marine ecosystems, such as whales, dolphins, octopi, and sharks, live in the open ocean.

As the depth of the ocean increases, it gets darker, colder, and with less available oxygen. Organisms living in deep-sea ecosystems within the dysphotic and aphotic zones have unusual adaptations that help them survive in these challenging environments. Some organisms have extremely large mouths that allow them to catch whatever nutrients fall from shallower ocean depths. Others have adapted to get their energy via chemosynthesis of chemicals from hydrothermal vents.

THE MARINE ENVIRONMENT

The shape of the oceans and seas of the world has changed significantly throughout the past 600 million years. According to the theory of plate tectonics, the crust of the Earth is made up of many dynamic plates. There are two types of plates—oceanic and continental—which float on the surface of the Earth's mantle, diverging, converging, or sliding against one another. When two plates diverge, magma from the mantle wells up and cools, forming new crust; when convergence occurs, one plate descends—i.e., is subducted—below the other and crust is resorbed into the mantle. Examples of both processes are observed in the marine environment. Oceanic crust is created along oceanic ridges or rift areas, which are vast undersea mountain ranges such as the Mid-Atlantic Ridge. Excess crust is reabsorbed along subduction zones, which usually are marked by deep-sea trenches such as the Kuril Trench off the coast of Japan.

The shape of the ocean also is altered as sea levels change. During ice ages a higher proportion of the waters of the Earth is bound in the polar ice caps, resulting in a relatively low sea level. When the polar ice caps melt during interglacial periods, the sea level rises. These changes in sea level cause great changes in the distribution of marine environments such as coral reefs. For example, during the last Pleistocene Ice Age the Great Barrier Reef did not exist as it does today; the continental shelf on which the reef now is found was above the high-tide mark.

Marine organisms are not distributed evenly throughout the oceans. Variations in characteristics of the marine environment create different habitats and influence what types of organisms will inhabit them. The availability of light, water depth, proximity to land, and topographic complexity all affect marine habitats.

The availability of light affects which organisms can inhabit a certain area of a marine ecosystem. The greater the depth of the water, the less light can penetrate until below a certain depth there is no light whatsoever. This area of inky darkness, which occupies the great bulk of the ocean, is called the aphotic zone. The illuminated region above it is called the photic zone, within which are distinguished the euphotic and disphotic zones.

The euphotic zone is the layer closer to the surface that receives enough light for photosynthesis to occur. Beneath lies the disphotic zone, which is illuminated but so poorly that rates of respiration exceed those of photosynthesis. The actual depth of these zones depends on local conditions of cloud cover, water turbidity, and ocean surface. In general, the euphotic zone can extend to depths of 80 to 100 metres and the disphotic zone to depths of 80 to 700 metres.

Marine organisms are particularly abundant in the photic zone, especially the euphotic portion; however, many organisms inhabit the aphotic zone and migrate vertically to the photic zone every night. Other organisms, such as the tripod fish and some species of sea cucumbers and brittle stars, remain in darkness all their lives.

Marine environments can be characterized broadly as a water, or pelagic, environment and a bottom, or benthic, environment. Within the pelagic environment the waters are divided into the neritic province, which includes the water above the continental shelf, and the oceanic province, which includes all the open waters beyond the continental shelf. The high nutrient levels of the neritic province—resulting from dissolved materials in riverine runoff—distinguish this province from the oceanic. The upper portion of both the neritic and oceanic waters—the epipelagic zone—is where photosynthesis occurs; it is roughly equivalent to the photic zone. Below this zone lie the mesopelagic, ranging between 200 and 1,000 metres, the bathypelagic, from 1,000 to 4,000 metres, and the abyssalpelagic, which encompasses the deepest parts of the oceans from 4,000 metres to the recesses of the deep-sea trenches.

The benthic environment also is divided into different zones. The supralittoral is above the high-tide mark and is usually not under water. The intertidal, or littoral, zone ranges from the high-tide mark (the maximum elevation of the tide) to the shallow, offshore waters. The sublittoral is the environment beyond the low-tide mark and is often used to refer to substrata of the continental shelf, which reaches depths of between 150 and 300 metres. Sediments of the continental shelf that influence marine organisms generally originate from the land, particularly in the form of riverine runoff, and include clay, silt, and sand. Beyond the continental shelf is the bathyal zone, which occurs at depths of 150 to 4,000 metres and includes the descending continental slope and rise. The abyssal zone (between 4,000 and 6,000 metres) represents a substantial portion of the oceans. The deepest region of the oceans (greater than 6,000 metres) is the hadal zone of the deep-sea trenches. Sediments of the deep sea primarily originate from a rain of dead marine organisms and their wastes.

WHAT ARE LIVING THINGS, AND HOW ARE THEY CLASSIFIED?

Classification is a way of categorizing organisms that enables an easy process of identification of such organisms. All living things are classified by grouping them under specific categories. The grouping is done based on the close relativeness of the organisms. The classification tool requires knowledge of characteristics, differences, and similarities between different organisms.

What are Marine Organisms?

Marine organisms are living entities that inhabit vast areas of seas and oceans. These organisms live in saline water bodies that support a wide range of lives. Marine organisms can be found floating on the surface as well as inhabiting the deep sea. Some marine plants carry out photosynthesis, while some creatures derive food from chemical substances where sunlight does not penetrate.

MARINE ORGANISMS CLASSIFICATION

Like all organisms, marine organisms are classified based on a hierarchical system. It includes the highest position kingdom, followed by phylum, class, order, family, genus, and species. Marine organisms can be classified by considering various factors, including habitat and mobility, metabolism, size, and developmental stages. These organisms are also scientifically named using genus and species. Genus is ranked above species in the hierarchy.

Types of Marine Organisms

Marine organisms are divided into several groups. These comprise several genera and species of bacterium, protozoans, planktons, marine invertebrates, and vertebrates. Based on habitat and mobility, there are three main groups of marine animals: plankton, nekton, and benthos. Based on the five-kingdom classification, there are numerous bacteria, protozoans, planktons, a very few fungi, plants, and diverse groups of animals. Based on the presence and absence of backbone, there are marine invertebrates and vertebrates. Different groups of marine organisms are:

Plankton

Planktons are considered floaters. They can be phytoplankton, zooplanktons, or bacterioplankton (plankton containing a bacterium's component). Phytoplankton is also called drifting algae, while zooplanktons are drifting protists and other minute animals. There are marine prokaryotes in a floating or attached state. These are mainly the archaea.

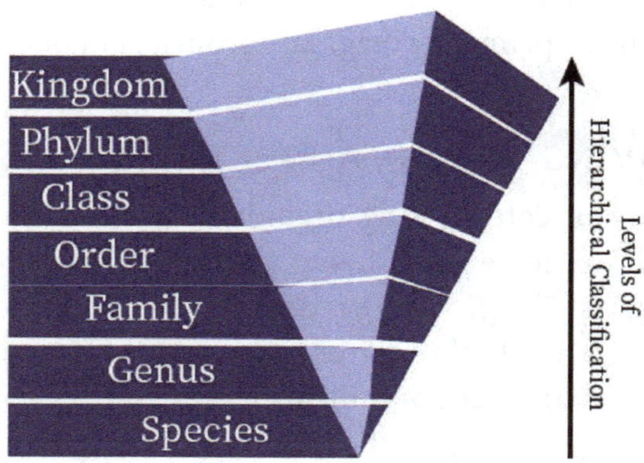

Fig. 1. Levels of hierarchical classification

Two types of zooplanktons exist, such as mesozooplankton and macro-zooplankton. Copepods are zooplanktons that filter out the phytoplankton present in the marine habitat. Macrozooplanktons are larger than copepods. They mostly serve as nutrition for whales.

Bacterioplankton includes all the bacteria that drift in the water. A bacterium called blue-green algae, which is a cyanobacterium, is the most common bacterioplankton. This bacterium is found in abundant quantities in the ocean. Examples of marine phytoplankton also include diatoms and dinoflagellates. Dinoflagellates are found in abundance, and most species are associated with producing bioluminescence.

Nekton

All swimmers of the marine habitat belong to the nekton community. These swimmers are found at all depths of the seas and oceans. Most of the marine vertebrates are swimmers. Fish are the most common creatures belonging to the nekton community. They are found in open water bodies and are also bottom dwellers. Based on their habitat, they are called pelagic fish and demersal fish.

The nekton community also includes mammals that spend their lives completely in the sea as well as animals that spend their lives both on the sea and on lands. Some examples include whales and seals, respectively.

Benthos

The benthos community comprises marine life, such as marine animals and plants that live in the benthic region of the marine habitat. The benthic region is the sea bottom that extends from the littoral zone to the open sea. Marine life is diverse in this region, and hence the benthic region has a huge variety of species. Out of all the known marine animals, 16% of them belong to the benthos community.

The benthos community is again subdivided into epi-flora and epifauna and in-flora and infauna. Epi-flora and epifauna inhabit the surfaces of sea bottoms. On the other hand, in-flora and infauna live partially or completely buried into the sea bottom. The benthos community feeds on organic matter and nekton animals. Examples of benthos animals include sea stars, several species of mollusks, and scavengers like hagfish.

Plants and Algae

Some plants, such as salt marsh grasses, are found in the saline habitat. Algae such as seaweed also inhabit the seas and oceans. True and false plants, such as salt marsh grasses and seaweeds, respectively, are primary producers. They are capable of photosynthesis owing to the region they are found. It is the neritic region where they can be abundantly observed, and sunlight penetrates this region.

Marine invertebrates

Diverse groups of invertebrates exist in the seas and oceans. Examples include jellyfish, starfish, sponges, and sea worms. Some of the marine invertebrates include Brachiopods, Phoronida, Nemertea, Cephalochordate, and Hemichordate.

- Brachiopoda are phyla of marine organisms with hard upper and lower shells.
- Phoronida is a phylum of filter feeders with soft bodies. They have upright chitin tubes for protection.
- Bryozoan is a phylum consisting of sedentary marine invertebrates. Such organisms live in colonies and are known as sea mats.
- Nemertea is another phylum of marine invertebrates. They comprise worms such as ribbon worms or proboscis worms.

- Porifera is a phylum that includes marine sponges. They are multicellular organisms with several pores and channels. They have cellular-level organization and lack body symmetry. The mode of nutrition is heterotrophic. They also possess spongin and spicules and lead a sedentary marine life.
- The cnidaria phylum of marine organisms has a tissue-level organization. The organisms under this phylum exhibit radial symmetry. The most common examples are sea anemones and jellyfish.
- Ctenophora are swimmers that possess cilia. The organisms under this phylum exhibit rotational symmetry. This phylum comprises organisms like comb jellies.
- Annelida consists of several worms and makes up a large phylum. Examples include burrowing polychaetes and sea leeches.
- Arthropoda phylum has more than 11 lakh species. These species have jointed appendages with strong moving and sensing characteristics. Crustaceans are marine arthropods, and examples include lobsters, crabs, and shrimps.
- Platyhelminthes is another phylum constituting worms. These marine tapeworms are found in the gut of large marine animals. An example includes a marine tapeworm called *Polygonoporus giganticus*.
- The Mollusca phylum consists of classes of organisms such as Gastropoda, Bivalvia, and Cephalopoda. It is the largest marine phylum. Examples include snails, octopuses, oysters, squid, and cuttlefish.
- Echinoderm phylum consists of organisms that belong to only marine habitats. They have a hard and spiny outer covering or skin. Most of the organisms are intertidal, and a few live on the ocean floor. Examples include sea cucumbers and sea urchins.

Other phyla belonging to marine invertebrates include Tardigrada and Lobopodia. The non-craniate phylum belonging to the marine invertebrates that are closely related to the marine vertebrates are as follows:

- A tunicate is a group of invertebrates found in the seas and oceans. They are found attached to the rocks or other surfaces of the ocean floor. Examples of tunicate include sea squirts and salps.
- Cephalochordata belongs to the phylum chordata but lacks a backbone. Amphioxus is an example of this group of invertebrates found in marine habitats.
- The phylum hemichordate includes deuterostome animals, such as acorn worms. Acorn worms are named as such because of their acorn-shaped front end.

Marine vertebrates

The marine vertebrates have a backbone and comprise the fish, reptiles, and mammals of the seas and oceans. These marine organisms are discussed below.

- Fish are ectothermic vertebrates that are also found in saltwater and lead marine life. They have a streamlined body apt for swimming. The appendages are fins that allow them to cut across the water. The body of a fish is covered with scales for protection. The respiration process occurs through the gills. The gases are exchanged through the gills, and oxygen is taken up from the water. Examples include sardines, sharks, and eels. Some of these marine fish also have swim bladders that help in buoyancy. Some fish also exhibit bioluminescence.
- Reptiles are also found in marine habitats. However, only a few types are found. Examples include sea turtles, snakes, and crocodiles. Apart from sea snakes, all these reptilian species lay eggs on land.
- Mammals are also found in the seas and oceans. They belong to several groups, such as cetaceans, sirenians, mustelids, and pinnipeds. Examples include whales, dolphins, sea lions, sea otters, seals, and walruses. All of these marine lives belonging to the class mammal are endothermic. They give direct birth to the young, are adapted to marine life such as swimming, and their kidneys excrete salt.

HOW ARE MARINE ORGANISMS CLASSIFIED?

A wide variety of organisms lives in the ocean, ranging in size from microscopic bacteria and algae to blue whales. All living things belong to one of the three major domains (branches) of life: Archaea, simple microscopic bacterialike creatures; Bacteria, simple life-forms consisting of cells that usually lack a nucleus; and Eukarya, complex organisms (including plants and animals) consisting of cells that have a nucleus.

Organisms are further divided into six kingdoms: Eubacteria, microscopic single-celled organisms without a nucleus; Archaebacteria, ancient bacterialike organisms that live in extreme environments; Plantae, many-celled plants; Animalia, many-celled animals; Fungi, molds and lichens, and Protista, single-celled and multicelled organisms with a nucleus. Classification of organisms involves placing individuals within the kingdoms into increasingly specific groupings of phylum, class, order, family, genus, and species, the last two of which denote an organism's scientific name. Many organisms also have one or more common names.

Marine organisms can be classified into one of three groups, based on habitat and mobility. Plankton are free-floating forms with little power of locomotion, nekton are swimmers, and benthos are bottom dwellers. Most of the ocean's biomass is planktonic.

HOW MANY MARINE SPECIES EXIST?

Only about 13% of all known species inhabit the ocean, and more than 98% of marine organisms are benthic. The marine environment—especially the pelagic environment—is much more stable than the terrestrial environment, so there is less pressure on marine organisms to diversify.

According to the World Register of Marine Species, WoRMS (World Register of Marine Species), the total number of marine species known to us is about 240,000 species (2021 census). However, it is estimated that there are 1.4 -1.6 million marine species on earth. Another, more recent estimate is about a half lower (0.7-1 million. The reason for the huge range in the estimated number of species is the lack of information on the diversity of some of the smaller organisms on Earth. For example, in the ocean, there is a plenty of information on marine mammals (seals, whales, dolphins, porpoises) and fish, while only recently are scientists beginning to understand the extreme diversity present in micro-organisms such as bacteria and phytoplankton (i.e. the plants of the sea). In Europe alone it is estimated that there are 41,000 - 56,000 species present (5,000 - 20,000 have yet to be identified!).

Since the WoRMS project began in 2008, it has also listed about 1,000 new types of fish - including a combined total of 122 sharks and rays, and a new barracuda in the Mediterranean sea. There are now about 18,000 known species of fish.

Marine life can have big economic value - sponges and molluscs are among species that have yielded cancer-fighting agents.

Mees, director of the Flanders Marine Institute in Belgium where WoRMS is based, said marine prospecting for "blue biotechnology" around volcanic vents on the seabed could also help develop materials resistant to heat and toxins.

Along with new species, a review by 200 editors also slashed about 190,000 species from the world lists after finding they duplicated already known organisms. That cut the total to 228,450 from almost 419,000.

MARINE ORGANISMS AND ADAPTATIONS

Adaptation is an evolutionary process whereby an organism becomes increasingly well suited to living in a particular habitat. It is not a quick process! Natural selection over many generations results in helpful traits becoming more common in a population. This occurs because individuals with these traits are better adapted to the environment and therefore more likely to survive and breed.

Adaptation is also a common term to describe these helpful or adaptive traits. In other words, an adaptation is a feature of an organism that enables it to live in a particular habitat.

DIFFERENT TYPES OF ADAPTATIONS

Marine organisms have adapted to the great diversity of habitats and distinctive environmental conditions in the marine environment. Adaptations are many and varied but they are generally grouped into 3 main categories: structural, physiological and behavioural.

Structural adaptations

Structural (or morphological) adaptations are the physical features of the organism. These include things you can see, like its shape or body covering, as well as its internal organisation. Following are a few of the ways that marine organisms have adapted their physical features to suit a particular habitat.

Seawater is much denser than air – as a result, there are vast numbers of microscopic organisms suspended in it. Cockles, as well as many other bivalves, are filter feeders. They have adapted specialised siphon structures to filter these organisms and any other particles of food from the surrounding water.

Estuaries have quite variable conditions – tides, waves and salinity fluctuations affect the animals and plants that live there on a daily basis. Many animals, such as cockles, are adapted to live in these conditions. They have strong shells that protect them from wave action, drying out and the prying beaks of predators.

Coastal plants need special adaptations to survive. For example, many types of seaweed attach firmly to rocks so they are not swept away by waves. Their leaf-like fronds are tough and leathery, which helps protect them from being torn by the waves or dried out by the sun.

Dolphins are mammals, but they look very different to mammals that live on land, as they are adapted to living in water. They have a streamlined shape and fins instead of legs. They also have blowholes on the tops of their heads. They use these to breathe, rather than through their mouths and noses. Find out more about New Zealand's two endemic dolphin species.

Physiological adaptations

Physiological adaptations relate to how the organism's metabolism works. These adaptations enable the organism to regulate their bodily functions, such as breathing and temperature, and perform special functions like excreting chemicals as a defence mechanism.

Some marine mammals, such as whales, migrate over large distances and may spend time in a combination of arctic, tropical and temperate waters. To cope with these temperature changes, they are endothermic or 'warm blooded'. This means that they are able to maintain a constant body temperature that is not dependent on the surrounding water.

Slow-moving species have adaptations that help protect them from predators. For example, many marine organisms can only move slowly or not all. This means they cannot easily get away from mobile predators, and they have other adaptations to protect them from being eaten. These can include chemical defences in their skin, for example, sea stars.

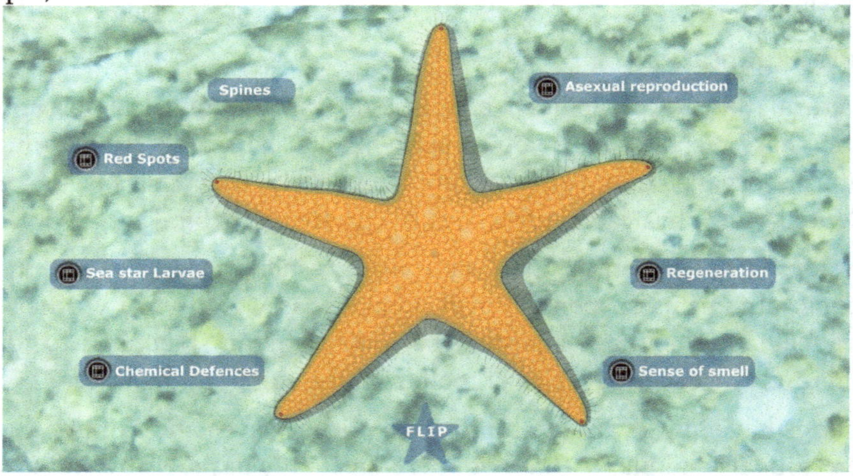

Fig. 2. Sea star adaptations – dorsal view. Sea stars have many weird and wonderful adaptations including both sexual and asexual reproduction.

Behavioural adaptations

Behavioural adaptations are learned or inherited behaviours that help organisms to survive, for example, the sounds made by whales allow them to communicate, navigate and hunt prey. Crab larvae use sounds to help them find suitable habitats so they can settle and metamorphose (change into an adult form).

Bryozoan colonies (Fig. 3) are found in high numbers on the continental shelf in New Zealand. They look like plants but are actually made up of hundreds of tiny individual animals that have banded together in order to more successfully find food and survive predation.

DIVISIONS OF MARINE ENVIRONMENT

Introduction

The ocean currently covers 71% of the earth's surface. Around two thirds of earth's land area is found in the Northern Hemisphere which means the ocean covers 61% of the total area, the rest 39% is land. However in the southern Hemisphere the ocean covers 80% of the area and the rest 20% is land. Marine ecosystem is the largest aquatic system of the planet which includes oceans, coral reefs, and estuaries. Since it is a single large and a complex system, it is very difficult to deal with it as a whole. Therefore the oceanographers have divided the ocean into many zones according to physical characteristics, mainly based on depth, light and temperature. The most basic division of the ocean based on location is between the water column and the bottom. The water portion is called the "Pelagic zone" (Greek pelagikos meaning sea) and the bottom is called the Benthic zone (Greek benthos meaning depths of sea). Each of these has subdivisions. Mariana Trench or Marianas Trench is the deepest natural trench in the world.

Fig. 3. Bryozoan colony

Photic or euphotic zone

The photic zone, euphotic zone (Greek for "well lit" or "light"), or sunlight (or sunlit) zone is the uppermost layer of water in a lake or ocean that is exposed to intense sunlight. It can also be measured as the depth where at least 1% light is available. This is the zone for primary production and respiration.

Aphotic zone

The aphotic zone (aphotic from Greek prefix "without light") is the portion of a lake or ocean where there is little or no sunlight. It is formally defined as the depths beyond which less than 1% of sunlight penetrates. Consequently, bioluminescence is essentially the only light found in this zone. Most food in this zone comes from dead organisms sinking to the bottom of the lake or ocean from overlying waters.

The depth of the aphotic zone can be greatly affected by such things as turbidity and the season of the year. The aphotic zone underlies the photic zone, which is that portion of a lake or ocean directly affected by sunlight. Depending on how aphotic zone is defined, the aphotic zone of the ocean begins between depths of roughly 200 m (660 ft) to 1,000 m (3,300 ft), and extends to the ocean floor. Temperatures can range from roughly 0 °C (32 °F) to 6 °C (43 °F. Unusual and unique creatures dwell in this expanse of pitch black water, such as the gulper eel, the giant squid, the anglerfish, and the vampire squid.

Pelagic realm

The pelagic realm or zone is divided into 2 horizontal zones- neritic and oceanic. The neritic zone is the water area between the low tide marks to the edge of the continental shelf. Instead of neritic many use the term continental shelf or coastal.

The oceanic zone is the open water area beyond that. The oce- anic zone is further divided into vertical regions called the epipelagic zone, mesopelagic zone, bathypelagic zone, abyssalpelagic zone, and hadalpelagic zone. These vertical regions can also be labeled as sunlit, twilight, midnight and abyssal (the very deep region).

Epipelagic zone

The epipelagic zone is the top layer that sunlight penetrates (Greek epi meaning over or before). It extends up-to 200m depth (all light rays are seen here initially). It is also called as the photic zone or euphotic zone. The epipelagic photic euphotic zone is the ideal place for about 90% of all ocean life to live because of warm temperature and sunlight .This is the only zone to support plant life because it has the light needed for photosynthesis.

As this region supports diverse plant life, variety of animals such as zooplankton, crustaceans, mollusks, sharks, sting rays, mackerels, tuna, seals, sea lions, sea turtles, etc., are abundant here. The majority of the epipelagic zone receives sun- light and lies in the photic zone - also simply called the sunlit zone. The photic zone is further divided between the euphotic (lighted) and dysphotic (dimly lit) zones. The aphotic zone is in constant dark- ness and comprise the seas below approximately 1,000 meters (3280 feet).

Mesopelagic zone (Greek mesos meaning middle)

It extends from 200 to1000 meters depth. It is also called as disphotic zone as; only blue light is seen here. It is also referred to as the "twilight zone"; its lower boundary in the tropics is the 10°C isotherm. "Heterotrophic bacteria" is dominant sp. in this zone.

Though some sunlight penetrates through this zone, it is not enough for photosynthesis. However this zone has octopus, squid etc. There animals tolerate cold temperatures, increased water pressure and darkness (like Squid and cuttlefish). Some fish have extra big eyes to help them see, while others produce their own light called bioluminescence using special organs in their bodies called photophores. Most fish don't chase their food but either wait for it or stalk it. Some have sharp fangs or big mouths to help them catch their food.

Bathypelagic zone (Greek bathos meaning deep):

This zone extends from 1000 up-to 4000 meters deep (aphotic zone; no light reaches this depth, there is total darkness). It lies between the boundaries of water with 10 and 4o C isotherm layers.

Abyssopelagic zone (Greek abyssos meaning bottomless)

It lies below 2000 and extends up-to 6000 meters depth (aphotic zone).

Hadalpelagic zone (Greek Hades meaning the underground abode of the dead)

It has a depth of 6000-10000 meters (aphotic zone). The Hadal zone is covers the deepest parts of the ocean. This zone is totally dark and cold, with intense pressure. Creatures found here have adapted themselves to the darkness by reducing the use of eyesight. Fishes occurring here do have eyes, and they are usually enormous, which indicates enough flashes of bioluminescent light to keep their eyes from totally deteriorating.

Collectively, these zones (bathypelagic, abyssalpelagic, and hadalpelagic) are simply called the "midnight" zone. The bathypelagic and hadalpelagic zones, do not have as many fish as the earlier zones. The coloration of the fish is black or red, and have bioluminescence (used to lure prey). The shape of the fishes living have is globular (round is shape and no streamlining). Most the weak swimmers live here and the fishes are mostly small but some are large. The eyes of the fishes are almost small or absent.

This zone is also called midnight zone or dark zone. This zone has a very intense water pressure which can be as great as two tons per square inch. Just like the mesopelagic zone, there are no plants and fewer animals which include vampire squid, giant squid, amphipod, slime stars, snake dragon fish, anglerfish, oarfish and gulper eel. The sperm whale dives to these depths search of food. Only about 1 % of all ocean species live in this zone, and some do not have eyes.

Benthic zone

The benthic zone is divided based on depth. Moving from shore toward the open ocean, the first zone is the supralittoral zone (Latin supra meaning upper and litus meaning shore). This is the zone that water splashes, but does not remain submerged. Beyond that lies the littoral zone. It is defined as the water area near the land or shore. Beyond the littoral zone is the continental shelf. This area is divided into the sublittoral zone, which is the ocean bottom close to shore, and the outer sublittoral zone, which is the ocean bottom out to the edge of the continental shelf.

CHAPTER 13

ANIMALS OF THE PELAGIC ENVIRONMENT

PELAGIC ZONE DEFINITION

"Pelagic zone is the region of the ocean outside the coastal areas and is also known as the open ocean."

INTRODUCTION

The pelagic zone is that region of lake, river or ocean that is not associated with the shore or the bottom. The pelagic zone organisms inhabiting the zone do not come in contact with the bottom or the shore throughout their lives. The pelagic zone is nutrient-poor. The large fish find their food by swimming long distances or drifting with currents and feeding on nutrient-deficient organisms.

Sub-Zones of Pelagic Zone

The pelagic zone is further divided into the following sub-zones based on the penetration of light.

Epipelagic Zone

- This zone stretches from the surface to a depth of about 200 metres.
- It is the surface zone where sufficient light penetrates for photosynthesis.
- This region is dominated by phytoplankton, diatoms, and dinoflagellates.
- Large fishes such as tunas and sharks are found in this zone. That is why the smaller animals come up to this zone only at night to stay away from the large predators around.

Mesopelagic Zone

- This zone extends from 200-1000 metres below the epipelagic zone.
- This is known as the twilight zone.
- Although some light reaches the region, it is not sufficient for photosynthesis.

- Some animals found in this zone have large eyes to make the best use of limited light.
- The oxygen concentration is also very low here.
- The organisms such as squids, nautilus shells, swordfish, etc. have the capacity to survive in low oxygen concentrations.

Bathypelagic Zone

- This is known as the dark zone where no light can reach.
- This zone extends from 1,000-4,000 metres below the continental slope.
- This is the zone of high pressure. The organisms with special features to withstand such high pressures can survive in this zone. E.g., the fishes found here lack a swimbladder.
- The organisms here stay still to conserve energy.
- The fishes exhibiting bioluminescence are prominent in this zone.
- The temperature here stays between 2-4°.
- Giant squids and sperm whales are found in this zone.

Abyssopelagic Zone

- This zone extends from 4,000- 6,000 metres and is the region where the continental slope levels off.
- It comprises more than 30% of the bottom ocean.
- The organisms here are colourless and blind.

Hadopelagic Zone

- This zone extends from 6,000-11,000 metres.
- Very few species are observed in this zone.
- Food availability is very low here.
- Many organisms found here live in hydrothermal vents.

The epipelagic and mesopelagic zones are the photic zones, while all the other zones are the aphotic zones with no light penetration.

PELAGIC WATERS

The Pelagic waters refer to open and free waters in the oceanic body. It stretches between the ocean surface and ocean bottom. Any marine life residing in the pelagic zone can swim freely without any boundary constraints.

Pelagic zone, an ecological realm that covers the entire ocean water column. Of the earth's inhabitants, the pelagic zone has the largest volume. It consists of the water column of the open ocean. It can be further divided into regions by depth. The Pelagic zone or water column can go from the surface of the sea to the entire bottom. As the water column changes with respect to the distance from the surface, its pressure increases, and in turn the temperature and light decrease. It results in a change of salinity, amount of dissolved oxygen, and micronutrients such as iron, magnesium, and calcium.

In addition to changes in the oceanic Pelagic zone, marine life is affected by underwater topography and by the surface of the seafloor or shoreline. Not only the discussed factors are responsible for marine life variations, instead, but the boundary between the ocean and atmosphere is also equally important. It brings the light for photosynthesis but can also bring predation from up and water currents.

PELAGIC ZONE ANIMALS

The organisms in the pelagic zone range from tiny planktons to large mammals like whales. Phytoplanktons provide oxygen for humans and food for many animals. Zooplanktons are also found in this zone. These are heterotrophic planktons that include most of the micro and macroorganisms.

Invertebrates like jellyfish, squids, octopus and krill are also found in the pelagic zone. Large ocean vertebrates such as crustaceans, sharks, bluefin tuna, and sea turtles live or migrate through the pelagic zone. Seabirds such as shearwaters, petrels and gannets can be found above the pelagic zone.

DEPTH AND LAYERS OF PELAGIC ZONE

Depending upon the depth of the sea, the pelagic zone can extend up to five vertical regions in the ocean. These are as follows:

1. **Epipelagic** - This is the lighter zone where enough light is present for photosynthesis. All primary production in the water occurs here. Plants and animals are concentrated in this layer. Examples of pelagic zone organisms residing here are plankton, seaweed, jellyfish, tuna, and dolphins.
2. **Mesopelagic** - The pelagic zone organisms residing in the mesopelagic zone are heterotrophic bacteria. Examples of organisms that live here are swordfish, squid, wolffish, and species of cuttlefish.
3. **Bathypelagic** - At this level of Pelagic water, the ocean is pitch black. No living plant exists here apart from occasional bioluminescent organisms such as anglerfish. Organisms living here survive by consuming detritus falling from the above zone. This phenomenon is also known as marine snow. The inhabitants living in this zone are giant squid, dumbo octopus.
4. **Abyssopelagic** - This zone starts from around 4000 meters. Very few creatures can reside in cold temperatures, high pressure, and complete

darkness. Species found in this zone are several species of squid, swimming cucumber, sea pig, and sea spider. Most of the species living here are transparent and eyeless. The reason being this evolution is the total lack of light.

5. **Hadopelagic -** This is the deepest part of the ocean lying within oceanic trenches. The zone is found from a depth of 6000 to 11000 meters. Most of the hadal habitat is found in the Pacific ocean.

PELAGIC ENVIRONMENT

The Pelagic environment is based on phytoplankton. They manufacture their own food using photosynthesis. They need sunlight, they inhabit the upper zone. The biodiversity reportedly decreases in the deeper zones. The pelagic zone organisms range from tiny planktons to large mammals. Phytoplanktons provide oxygen for humans and food for many organisms.

Zooplanktons are found in this zone. It includes heterotrophic planktons consisting of most of the micro and microorganisms.

Invertebrates like squids, jellyfish, octopus, and krill are found in the pelagic zone. Large ocean vertebrates such as crustaceans, sharks, bluefin tuna, sea turtles migrate through the pelagic zone. Pelagic birds live on open seas rather than inland. They feed on planktonic crustaceans, squids, and forage fish. The Pelagic environment consists of the following:

1. **Pelagic Invertebrates -** Low latitudes invertebrates tend to produce a large number of eggs whereas high latitude invertebrates produce fewer and larger eggs and larger offsprings.
2. **Pelagic Fish -** These live in the coastal ocean and lake waters but not on or near the bottom of the sea. These fishes are migratory forage fish. They feed on planktons.
3. **Pelagic Reptiles -** The pelagic sea snake is the only reptile of one of the 65 species of marine snake to spend its entire life in the pelagic zone. It is helpless on land. Many species of sea turtles spend the first year of life in the pelagic zone and on reaching maturity they move closer to the shore.

CHALLENGES IN THE PELAGIC ZONE

The species in this zone might be affected by the wave and wind activity, pressure, availability of prey, and temperature of the water. The pelagic zone is spread over a wide area. The prey might be scattered over some distance and the animals might have to travel very far in search of it.

Some animals have to travel miles between the breeding and feeding grounds. They encounter changes in the water temperature, human activities and types of prey. Thus, the animals in the pelagic zone face a few challenges.

CHAPTER

ANIMALS OF THE BENTHIC ENVIRONMENT 14

INTRODUCTION

The benthic zone begins at the shore and extends to the bottom of a waterbody. It is found all over the world. Only a small amount of sunlight reaches this zone. This zone is characterized by low temperature and high pressure. Such conditions are not optimum for sustaining vast flora and fauna found in this zone. The sediment layers of the benthic zone help in recycling the nutrients that helps in the survival of the aquatic life in the upper column.

The benthic zone refers to the ecological zone located at the bottom of any marine or freshwater body, such as a river, ocean, lake or pond. The benthic zone includes the sediment surface. Benthic zones are found all across the world, wherever there is an appreciable water body.

This implies that the benthic zone could be as shallow as a few centimeters, but may reach a depth of a few thousand meters. On account of the depths it can reach, the benthic zone is usually characterized by low temperature, high pressure and minimal sunlight.

Well, such conditions aren't optimal for the sustenance of a vast amount of flora and fauna, but sediment layers in the benthic zone help in the recycling of nutrients, which in turn makes it possible for the sustenance of rich aquatic life in the upper water column. The benthic zone might not be brimming with vivid aquatic creatures, but many crustaceans, snails, sponges, sea stars etc. are found here.

Pressure

Just like temperature, pressure also varies across the benthic zone. From being low for shallow benthic zones to very high for those located a few thousand meters deep. The Mariana Trench, which happens to be the deepest trench on Earth (around 10,000 meters below sea level), is the extreme example for a very high-pressure benthic zone. Here, the pressure is roughly 1000 times the normal pressure (sea level).

High-pressure benthic zones lead to a very homogenous environment, which produces organisms with distinctive traits. One peculiar example is the enlarged sizing of certain marine creatures. Also, at great oceanic depth, the quantity of dissolved oxygen in the water is higher, which makes organisms grow larger than their counterparts in more temperate zones.

Temperature

For benthic zones located close to the shore or with notably shallow depth, the temperature is warmer, but for those located hundreds of meters deep, temperatures can fall to 2°C to 3°C (near the abyssal zone). Aquatic life that can survive in such extreme cold temperatures is limited and those found there often move at a slow pace to conserve energy.

Light

The intensity of light decreases with depth. It becomes limited at depths below 250 meters; between 250 and 1000 meters, the intensity of light begins to dissipate very quickly. This zone is called the dysphotic zone and performing photosynthesis here is very difficult. Beyond a depth of 1000 meters, light availability is almost zero, with photosynthesis no longer being feasible.

BENTHOS

Because the benthic zone can occur in varying environments, chemical and physical characteristics vary greatly and are often dependent on context. Nutrient availability in the deep sea is also scarce, so organisms need to adapt themselves accordingly to survive here.

Organisms living in the benthic zone are called **benthos**. Benthos have specially adapted themselves to live on the bottom substrate in deep-water bodies with elevated pressure and cold temperatures. In fact, organisms that inhabit the deep-water pressure areas cannot survive in the upper parts of the water column. Most of these benthos are detritivores.

MOLLUSCA

Mollusca (or **mollusks**) are a very diverse groups of animal with at about 85,000 living species. Mollusks are the largest marine phylum, comprising about 23% of all known marine organisms. Mollusks include clams, scallops, oysters, mussels, limpets, chitons, and snails (snails are gastropods—the account for about 80% of invertebrate species). **Cephalopods** are mollusks and include octopuses, squid, cuttlefish, and nautilus.

- Mollusks all have unsegmented soft bodies with a "head" and a "foot" region (they may not look like a head or foot!).

- Often their bodies are covered by a hard exoskeleton, as in the shells of snails and clams or the plates of chitons.
- Many have shells, either calcareous, or made of proteins and chitin.
- Most mollusks have eyes.
- Mollusks have a mantle with a body cavity (used for breathing and excretion), and the presence of a radula (something tongue-like).
- All mollusks larvae nervous system, blood circulation system, and often complex digestive system.
- All produce eggs that emerge as larvae or miniature adults.

Mollusks appeared in the Cambrian Period and have diversified into their multiple forms. A large group called **ammonites** dominated the oceans during the Mesozoic era, but vanished with many other species at the K/T Boundary extinction event. Their distant relatives, **squids**, that do not have calcareous shells, survived the K/T extinction event. Another distant relative, the **nautilus**, also survived the K/T event.

Fig. 1. Scallop

Fig. 2. Nautilus

GLOSSARY

Abyssal plain. The deep ocean floor, an expanse of low relief at depths of 4,000 to 6000 m

Abyssobenthic. Referring to benthic organisms living at abyssal depths.

Abyssopelagic zone. The 4000 to 6000-m-depth zone, seaward of the shelf-slope break

Acclimation. Given a change of a single parameter, a readjustment of the physiology of an organism, reaching a new steady state

Accuracy. Is the correctness of a measure when comparing to a known standard. See also resolution and precision.

Action spectrum. Portion of the light spectrum used in photosynthesis.

Active suspension feeders. Suspension feeders that produce a current to actively entrain and capture particles.

Aerosol. A variety of substances that may seed clouds and therefore increase cloud cover.

Age structure. The relative abundance of different age classes in a population

Aggregated spatial distribution. A case where individuals in a space occur in clusters too dense to be explained by chance

Ahermatypic. Non-reef-building (referring to scleractinian corals)

Allee effect. The increase of probability of extinction when the population is so small that population density itself matters. Finding mates in a sparse population is the major mechanism of decline in such a case.

Allele. One of several variants that can occupy a locus on a chromosome

Allopatric speciation. The differentiation of geographically isolated populations into distinct species

Allozyme. A variant of an enzyme type. These may be variants of a specific enzyme (e.g., cytochrome c) that are the products of a single genetic locus.

Amensal. Negatively affecting one or several species

Amino acids. Basic structural unit of proteins

Amnesic shellfish poisoning. Uptake by shellfish of domoic acid when feeding on certain species of Pseudo-nitzschia, resulting in being poisonous for consumption by humans.

Anadromous fish. Fish that spends most of its life feeding in the open ocean but that migrates to spawn in fresh water

Anaerobes. Organisms that carry out metabolic processes in the absence of oxygen. See also Aerobes.

Anoxic. Lacking oxygen.

Antibody. A specialized protein produced in response to the introduction of a foreign molecule into the body, usually to destroy the foreign body.

Aposematic coloration. A conspicuous coloration borne by an organism that has evolved to signal to a predator the organism's distastefulness or even poisonous state.

Applied marine biology. The use of marine biology to study practical problems such as pollution, or the use of biological structures to protect against wave surge.

Aragonite. An orthorhombic crystal variant of calcium carbonate that is less stable and more soluble than another variant, calcite. A common constituent of coral skeletons and many snail shells (exception: genus Conus).

Aragonite saturation horizon (ASH). Depth below which aragonite-secreting organisms cannot secrete aragonite and live.

Arctic Oscillation. Temporal variability of climate involving winds circulating counterclockwise around the Arctic at around 55° N latitude.

Arrow worms. Members of the phylum Chaetognatha, a group of planktonic carnivores

Asexual reproduction. Reproduction of the individual without the production of gametes and zygotes

Assimilation efficiency. The fraction of ingested food that is absorbed and used in metabolism

Assortative mating. The mating of a given genotype mates with another genotype at a frequency disproportionate to that expected from random encounter

Atoll. A horseshoe or circular array of islands, capping a coral reef system perched around an oceanic volcanic seamount

Attenuation (of light). Diminution of light intensity; explained, in the ocean, in terms of absorption and scattering

Autotrophic algae. Algae capable of photosynthesis and growth using only dissolved inorganic nutrients

Auxotrophic algae. Algae requiring a few organically derived substances, such as vitamins, along with dissolved inorganic nutrients for photosynthesis

Bathypelagic zone. The 2,000 to 4,000-m-depth zone seaward of the shelf-slope break

Benthic-pelagic coupling. The cycling of nutrients between the bottom sediments and overyling water column

Benthos. Organisms that live associated with the sea bottom. Examples include burrowing clams, sea grasses, sea urchins, acorn barnacles.

Berm. A broad area of low relief in the upper part of a beach

Between-habitat comparison. A contrast of diversity in two localities of differing habitat type (e.g., sand versus mud bottoms)

Biogenically reworked zone. The depth zone, within a sediment, that is actively burrowed by benthic organisms

Biogenic graded bedding. A regular change of sediment median grain size with depth below the sediment-water interface caused by the activities of burrowing organisms

Bioluminescence. Light emission, often as flashes, by many marine organisms

Blood pigment. A molecule used by an organism to transport oxygen efficiently, usually in a circulatory system (e.g., hemoglobin)

Bloom. (phytoplankton) A population burst of phytoplankton that remains within a defined part of the water column

Bohr effect. When blood pH decreases, the ability of hemoglobin to bind to oxygen decreases. An adaptation to release oxygen in the oxygen starved tissues in capillaries where respiratory carbon dioxide lowers blood pH

Boreal. Pertaining to the Northern Hemisphere, north temperate zone

Boring. Capable of penetrating a solid substratum by scraping or chemical dissolution

Bottom-up control. Refers to food webs. A control of a population that comes from change lower in a food web (e.g., control of a population of mussels by abundance of phytoplankton food).

Boundary layer. A layer of fluid near a surface, where flow is affected by viscous properties of the fluid. At the surface, fluid velocity must be zero, and the boundary layer is a thin film that depends on surface texture, fluid velocity in the "mainstream of flow," and fluid mass properties such as salinity.

Brackish sea. Semienclosed water body of large extent in which tidal stirring and seaward flow of freshwater do not exert enough of a mixing effect to prevent the body of water from having its own internal circulation pattern

Browsers. Organisms that feed by scraping thin layers of living organisms from the surface of the substratum (e.g., periwinkles feeding on rock-surface diatom films; urchins scraping a thin, filmy sponge colony from a rock)

Calcareous. Made of calcium carbonate

Carrying capacity. The total number of individuals of a population that a given environment can sustain

Carnivore An organism that captures and consumes animals

Catadromous fish. Fish that spawns in seawater but feed and spends most of its life in estuarine or fresh water

Character displacement. A pattern in which two species with overlapping ecological requirements differ more when they co-occur than when they do not. The difference is usually in a morphological feature related to resource exploitation, as in the case of head size, which may be related to prey size

Chemosynthesis. Primary production of organic matter, using various substances instead of light as an energy source; confined to a few groups of microorganisms

Chlorinity. Grams of chloride ions per 1000 grams of seawater

Chloroplast. In eukaryotic organisms, the cellular organelle in which photosynthesis takes place

Cladogram. A tree-like diagram showing evolutionary relationships. Any two branch tips sharing the same immediate node are most closely related. All taxa that can be traced directly to one node (that is they are "upstream of a node") are said to be members of a monophyletic group.

Coastal reef. A coral reef occurring near and parallel to a coastline

Comb jellies. Members of the phylum Ctenophora, a group of gelatinous forms feeding on smaller zooplankton

Commensal. Having benefit for one member of a two-species association but neither positive nor negative effect on the other

Compensation depth. The depth of the compensation light intensity

Compensation light intensity. That light intensity at which oxygen evolved from a photosynthesizing organism equals that consumed in its respiration

Competition. An interaction between or among two or more individuals or species in which exploitation of resources by one affects any others negatively

Complex life cycle. A life cycle that consists of several distinct stages (e.g., larva and adult)

Conformer. An organism whose physiological state (e.g., body temperature) is identical to, and varies identically with, that of the external environment

Continental drift. Horizontal movement of continents located in plates moving via sea-floor spreading

Continental shelf. A broad expanse of ocean bottom sloping gently and seaward from the shoreline to the shelf-slope break at a depth of 100 to 200 m

Convergence. The contact at the sea surface between two water masses converging, one plunging below the other

Copepod. Order of crustaceans found often in the plankton

Coprophagy. Feeding on fecal material

Coral reef. A wave-resistant structure resulting from cementation processes and the skeletal construction of hermatypic corals, calcareous algae, and other calcium carbonate-secreting organisms

Corer. Tubular benthic sampling device that is plunged into the bottom in order to obtain a vertically oriented cylindrical sample

Coriolis effect. The deflection of air or water bodies, relative to the solid earth beneath, as a result of the earth's eastward rotation

Countercurrent exchange mechanism. Mechanism by which two vessels are set side by side, with fluid flowing in opposite directions, allowing efficient uptake and retention of heat, oxygen, or gas, depending upon the type of exchanger

Countershading. Condition of organisms in the water column that are dark-colored on top but light-colored on the bottom

Counter-illumination. Having bioluminescent organs that are concentrated on the ventral surface so as to increase the effect of countershading (see also countershading)

Critical depth. That depth above which total integrated photosynthetic rate equals total integrated respiration of photosynthesizers

Critical salinity. A salinity of approximately 5 to 8% that marks a minimum of species richness in an estuarine system

Daily estuary. An estuary in which tidal movements cause substantial changes in salinity at any one location on a daily basis

Deep layer. The layer extending from the lowest part of the thermocline to the bottom

Deep-scattering layer. Well-defined horizon in the ocean that reflects sonar; indicates a layer usually consisting of fishes, squid, or other larger zooplankton

Demographic. Referring to numerical characteristics of a population (e.g., population size, age structure)

Density. (seawater) Grams of sea water per milliliter of fluid

Density-dependent factors. Factors, such as resource availability, that vary with population density

Deposit feeder. An organism that derives its nutrition by consuming some fraction of a soft sediment

Detritus. Particulate material that enters into a marine or aquatic system. If derived from decaying organic matter it is organic detritus.

Diatom. Dominant planktonic algal form with siliceous test, occurring as a single cell or as a chain of cells

Diffusion. The net movement of units of a substance from areas of higher concentration to areas of lower concentration of that substance

Digestion efficiency. The fraction of living food that does not survive passage through a predator's gut

Dinoflagellate. Dominant planktonic algal form, occurring as a single cell, often biflagellate

Directional selection. Preferential change in a population, favoring the increase in frequency of one allele over another

Dissolved organic matter. Dissolved molecules derived from degradation of dead organisms or excretion of molecules synthesized by organisms

Disturbance. A rapid change in an environment that greatly alters a previously persistent biological community

Diversity. A parameter describing, in combination, the species richness and evenness of a collection of species. Diversity is often used as a synonym for species richness

Diversity gradient. A regular change in diversity correlated with a geographic space or gradient of some environmental factor

Ekman circulation. Movement of surface water at an angle from the wind, as a result of the Coriolis effect

El Niño-Southern Oscillation (ENSO). Condition in which warm surface water moves into the eastern Pacific, collapsing upwelling and increasing surface-water temperatures and precipitation along the west coast of North and South America

Emigration. The departure of individuals from a given area

Endosymbiotic. Being symbiotic and living within the body of an individual of the associated species

Environmental stress. Variously defined as (a) an environmental change to which an organism cannot acclimate and (b) an environmental change that increases the probability of death

Epibenthic (epifaunal or epifloral). Living on the surface of the bottom

Epidemic spawning. Simultaneous shedding of gametes by a large number of individuals

Epipelagic zone. The 0- to 150-m-depth zone, seaward of the shelf-slope break

Epiphyte. Microalgal organism living on a surface (e.g., on a seaweed frond)

Estuarine flow. Seaward flow of low-salinity surface water over a deeper and higher -alinity layer

Estuarine realms. Large coastal water regions that have geographic continuity,are bounded landward by a stretch of coastline with fresh-water input, and are bounded seaward by a salinity front

Estuary. A semienclosed body of water that has a free connection with the open sea and within which seawater is diluted measurably with freshwater that is derived from land drainage

Euphausiid. Member of an order of holoplanktonic crustacea

Eutrophic. Water bodies or habitats having high concentrations of nutrients

Evenness. The component of diversity accounting for the degree to which all species are equal in abundance, as opposed to strong dominance by one or a few species

Fecal coliform bacteria. Technically, all the facultative anaerobic gram negative non-spore forming rod shaped bacteria that fermet lactose in EC medium with gas production within 24h at 44.5 degrees C. A measure of bacteria mostly originating from guts that enter waters. Believed to be correlated with disease-causing (pathogenic) bacteria.

Fecundity. The number of eggs produced per female per unit time (often: per spawning season)

Foliose coral. A coral whose skeletal form approximates that of a broad, flattened plate

Food chain. An abstraction describing the network of feeding relationships in a community as a series of links of trophic levels, such as primary producers, herbivores, and primary carnivores

Food chain efficiency. Amount of energy of some other quantity extracted from a trophic level, divided by the amount of energy produced by the next-lower trophic level

Food web. A network describing the feeding interactions of the species in an area

Foraminifera. Protozoan group, individuals of which usually secrete a calcareous test; both planktonic and benthic representatives

Founder principle. A small colonizing population is genetically unrepresentative of the source of population

Freshet. An increase of water flow into an estuary during the late winter or spring, owing to increased precipitation and snow melt in the watershed

Front. A major discontinuity separating ocean currents and water masses in any combination

Fugitive species. A species adapted to colonize newly disturbed habitats

Gametophyte. Haploid stage in the life cycle of a plant

Generation time. The time period from birth to average age of reproduction

Genetic drift. Changes in allele frequencies that can be ascribed to random effects

Genetic locus. A location on a chromosome (possibly of a diploid organism with variants that segregate according to the rules of Mendelian heredity)

Genetic polymorphism. Presence of several genetically controlled variants in a population

Genotype. The genetic makeup of an organism, with respect to a given genetic locus, the alleles it carries

Genus. (plural: genera) The level of the taxonomic hierarchy above the species but below the family level

Geostrophic flow. Movement of water in the oceans as a combined response to the Coriolis effect and gravitational forces created by an uneven sea surface

Geotactic. Moving in response to the earth's gravitational field

GIS. Geographic Information System. A system that allows automatic location of information suitable for mapping. Usually involves a software system that takes geographic position data and other data (e.g., type of bottom sediment) in order to create a map. Data on processes (e.g., current speed) can be incorporated to make a geographic model of flow.

Global warming. Predicted increase in the earth's oceanic and atmospheric temperature, owing to additions of carbon dioxide to the atmosphere, as a result of human activities

GPS. Global Positioning System. An electronic device that uses positioning signals from satellites in order to locate precisely latitutude and longitude. Now used nearly exclusively for locating ship sampling stations at sea, but also useful for locations near and on shore.

Grab. Benthic sampling device with two or more curved metal plates designed to converge when the sampler hits bottom and grab a specified volume of bottom sediment

Grazer. A predator that consumes organisms far smaller than itself (e.g., copepods graze on diatoms)

Greenhouse effect. Carbon dioxide traps solar-derived heat in the atmosphere near the earth.

Gregarious settling. Settlement of larvae that have been attracted to members of their own species

Gross primary productivity. The total primary production, not counting the loss in respiration

Guild. A group of species, possibly unrelated taxonomically, that exploit overlapping resources

Gyre. Major cyclonic surface current systems in the oceans

Halocline. Depth zone within which salinity changes maximally

Hardy-Weinberg law. Law that states that the frequencies of genotypes in a population at a locus are determined by random mating and allele frequency

Herbivore. An organism that consumes plants

Harmful algal bloom. A bloom of (usually) planktonic microalgae belonging to a strain of a species that has a toxic harmful to marine organisms or humans consuming marine organisms.

Heritable character. A morphological character whose given state can be explained partially in terms of the genotype of the individual

Hermaphrodite. An individual capable of producing both eggs and sperm during its lifetime

Hermatypic. Reef-building

Heterotrophic algae. Algae that take up organic molecules as a primary source of nutrition

Heterozygote. With respect to a given genetic locus, a diploid individual carrying two different alleles

Highly stratified estuary. An estuary having a distinct surface layer of fresh or very-low-salinity water, capping a deeper layer of higher salinity, more oceanic water

Histogram. A multiple-bar diagram representing the frequency distribution of a group as a function of some variable. The frequency of each class is proportional to the length of its associated bar

Holoplankton. Organisms spending all their life in the water column and not on or in the sea bed

Homeotherm. An organism that regulates its body temperature despite changes in the external environmental temperature

Homozygote. With respect to a given genetic locus, a diploid individual carrying two identical alleles

Hydrographic. Referring to the arrangement and movement of bodies of water, such as currents and water masses

Hydrothermal vents. Sites in the deep ocean floor where hot, sulfur-rich water is released from geothermally heated rock

Hypothesis. A refutable statement about one or a series of phenomena

Infaunal. Living within a soft sediment and being large enough to displace sedimentary grains

Interspecific competition. Condition in which one species' exploitation of a limiting resource negatively affects another species

Interstitial. Living in the pore spaces among sedimentary grains in a soft sediment

Isotonic. Having the same overall concentration of dissolved substances as a given reference solution

Keystone species. A predator at the top of a food web, or discrete subweb, capable of consuming organisms of more than one trophic level beneath it

Laminar flow. The movement of a fluid where movement of the entire fluid is regular and with parallel streamlines.

Larva. A discrete stage in many species, beginning with zygote formation and ending with metamorphosis

Larvacea. A group of planktonic tunicates that secrete a gelatinous house, used to strain unsuitable particles (large particles are rejected). An inner filter apparatus of the house, the so-called food trap or particle-collecting apparatus, is used to retain food particles.

LD50 The value of a given experimental variable required to cause 50% mortality.

Leaching. The loss of soluble material from decaying organisms

Lecithotrophic larva. A planktonic-dispersing larva that lives off yolk supplied via the egg

Leeward. The side of an island opposite from the one facing a persistent wind

Life table. A table summarizing statistics of a population, such as survival and reproduction, all broken down according to age classes

Litter. Accumulations of dead leaves in various states of fragmentation and decomposition

Logistic population growth. Population growth that is modulated by the population size relative to carrying capacity. Population growth declines as population approaches carrying capacity, and is negative when population size is greater than carrying capacity

Longshore current. A current moving parallel to a shoreline

Macrobenthos. (macrofauna or macroflora) Benthic organisms (animals or plants) whose shortest dimension is greater than or equal to 0.5 mm

Macrofauna. Animals whose shortest dimension is greater than or equal to 0.5 mm

Macrophyte. An individual alga large enough to be seen easily with the unaided eye

Macroplankton. Planktonic organisms that are 200-2,000 micrometers in size.

Mainstream flow. The flow in a part of the fluid (e.g., in a tidal creek) that is well above the bottom or well away from a surface and essentially not under the influence of the boundary layer.

Mangel. See Mangrove forest.

Mangrove forest. A shoreline ecosystem dominated by mangrove trees, with associated mud flats.

Marine snow. Fragile organic aggregates, resulting from the collision of dissolved organic molecules or from the degradation of gelatinous substances such as larvacean houses. Usually enriched with microorganisms.

Marine protected area. A conservation geographic unit designed to protect crucial communities and to provide reproductive reserves for fisheries that hopefully will disperse over wider areas.

Maximum sustainable yield. In fisheries biology, the maximum catch obtainable per unit time under the appropriate fishing rate

Meiobenthos (meiofauna or meioflora). Benthic organisms (animals or plants) whose shortest dimension is less than 0.5 mm but greater than or equal to 0.1 mm

Meiofauna. Animals whose shortest dimension is less than 0.5 mm but greater than or equal to 0.1 mm

Megaplankton. Planktonic organisms that are greater than or equal to 2000 micrometers in size

Meroplankton. Organisms that spend part of their time in the plankton but also spend time in the benthos (e.g., planktonic larvae of benthic invertebrates)

Mesopelagic. The 150 — 2000 m depth zone, seaward of the shelf-slope break

Metabolic rate. The overall rate of biochemical reactions in an organism. Often estimated by rate of oxygen consumption in aerobes

Metamorphosis. Major developmental change as the larva develops into an immature adult

Metapopulation. A group of interconnected subpopulations, usually of subequal size. The features of individuals now found in one subpopulation might have been determined by conditions affecting them when they were located in another subpopulation.

Microbenthos. (microfauna or microflora) Benthic organisms (animals or plants) whose shortest dimension is less than 0.1 mm

Microfauna. Animals whose shortest dimension is less than 0.1 mm

Mixing depth. The water depth to which wind energy evenly mixes the water column

Mixoplankton. Planktonic organisms that can be classified at several trophic levels. For example, some ciliates can be photosynthetic but also can ingest other plankton and are heterotrophic.

Moderately stratified estuary. An estuary in which seaward flow of surface low-salinity water and moderate vertical mixing result in a modest vertical salinity gradient

Monophyletic. Refers to a group of species that all have a single common ancestral species

Mucous-bag suspension feeder. Suspension feeder employing a sheet or bag of mucus to trap particles nonselectively

Mutualism. An interaction between two species in which both derive some benefit

Mutualistic. Conferring reciprocal benefit to individuals of two different associated species

Nanoplankton. Planktonic organisms that are 2-20 micometers in size.

Neap tides. Tides occurring when the vertical range is minimal

Nekton. Organisms with swimming abilities that permit them to move actively through the water column and to move against currents

Neritic. Seawater environments landward of the shelf-slope break

Net primary productivity. Total primary production, minus the amount consumed in respiration

Neuston. Planktonic organisms associated with the air-water interface.

Niche. A general term referring to the range of environmental space occupied by a species

Niche overlap. An overlap in resource requirements by two species

Nitrogen fixation. The conversion of gaseous nitrogen to nitrate by specialized bacteria.

North Atlantic Oscillation (NAO). A cycle of changing difference in air pressure between a low atmospheric pressure over Iceland at about 64° N latitude and a higher atmospheric pressure over the Azores, at approximately 38° N latitude.

North Pacific Transition Zone. A belt across the Pacific centered at about 35–40° N latitude where frontal transitions often result in high primary productivity.

No-slip condition. The condition where water has zero velocity when in contact with a surface.

No-take Reserves. Geographic areas where by law no one is allowed to fish or collect biological specimens. Rules could apply to one or all species.

Nuisance bloom. A rapid increase of one or only a few species of phytoplankton, resulting in densities high enough to cause discoloration of the surface water, possible increase of toxins, and degradation of water quality aspects such as dissolved oxygen

Nutrient cycling. The pattern of transfer of nutrients between the components of a food web

Nutrients. Those constituents required by organisms for maintenance and growth (we use this term in this book in application to plants)

Ocean acidification. The decline of pH in the ocean, owing to additions of carbon dioxide in the atmosphere and subsequent dissolution in seawater.

Oceanic. Associated with sea-water environments seaward of the shelf-slope break

Oceanic boundary currents. A distinct ocean current that runs parallel to and offshore of the coast usually on the western sides of oceans (e.g., the Gulf Stream).

Oceanic ridge. A sinuous ridge rising from the deep-sea floor

Oligotrophic. Refers to water bodies or habitats with low concentrations of nutrients

Omnivory. Being able to feed in more than one distinct way (e.g., an organism capable of carnivory and herbivory)

Optimal foraging theory. A theory designed to predict the foraging behavior that maximizes food intake per unit time

Organic. Deriving from living organisms

Organic osmolytes. Compounds including amino acids and urea that are used in osmoregulation.

Organic nutrients. Nutrients in the form of molecules synthesized by or originating from other organisms

Oscillation, climate. A fluctuation in air pressure characteristics, wind systems, sea-surface temperature (SST), or other weather features that occur on the geographic scale of an ocean, such as throughout the North Atlantic Ocean or perhaps even the world ocean.

Osmoconformer. An organism whose body fluids change directly with a change in the concentrations of dissolved ions in the external medium

Osmoregulator. An organism that regulates the concentration of dissolved ions in its body fluids irrespective of changes in the external medium

Osmosis. The movement of pure water across a membrane from a compartment with relatively low dissolved ions to a compartment with higher concentrations of dissolved ions

Otolith. A small mass of calcium carbonate, used in the inner ear of a fish for perception of balance. The mass often grows in increments that can allow us to determine the age of the fish.

Outwelling. The outflow of nutrients from an estuary or salt-marsh system to shelf waters

Overdominance. Selection favoring heterozygotes

Overfishing. Case in which high degree of fishing results in decline of a fish population to unsustainable levels.

Oviparous. Reproduction in which eggs are laid and hatch outside the mother's body, with little or no development within the mother. An example would be an unfertilized egg shed into the water that is fertilized by planktonic sperm.

Ovoviparous. Fertilization is internal, but eggs develop within the mother's body with no nutrients supplied. Eggs are eventually released into the environment and hatch.

Oxygen dissociation curve. A curve showing the percent saturation of a blood pigment, such as hemoglobin, as a function of oxygen concentration of the fluid

Oxygen minimum layer. A depth zone, usually below the thermocline, in which dissolved oxygen is minimal

Oxygen technique (primary productivity). The estimation of primary productivity by the measurement of the rate of oxygen increase

Pacific Decadal Oscillation. A climatic oscillation in the Pacific of a period of 20–30 years, in which cool and warm waters exchange locations in the eastern and western Pacific north of 20° N latitude.

Paradox of the plankton. The coexistence of many species of phytoplankton, despite evidence of resource limitation.

Paralytic shellfish poisoning (PSP). The ingestion of dinoflagellates containing saxitoxin by bivalves, which in turn may be eaten by humans, who are poisoned.

Parapatric speciation. The differentiation into distinct species of populations experiencing some gene flow

Parasite. An organism living on or in, and negatively affecting, another organism.

Partially mixed estuary. An estuary or area of an estuary with a gradient of lower-salinity to higher-salinity water as one moves from the surface to the bottom; mixed vertically by wind and tidal motion.

Particulate organic matter. Particulate material in the sea derived from the decomposition of the nonmineral constituents of living organisms

Patchiness. A condition in which organisms occur in aggregations

Pelagic. Living in the water column seaward of the shelf-slope break

Pellets. Compacted aggregations of particles resulting either from egestion (fecal pellets) or from burrow-constructing activities of marine organisms

Penetration anchor. In hydraulically burrowing organisms, any device used to penetrate and gain an initial purchase on the sediment so that the body can be thrust in farther

Peptides. Chains of amino acids; often portions of a protein molecule

pH. Measure of he acidity or basicity of water (-log10 of the activity of hydrogen ions in water)

Phenotypic plasticity. The capacity of an individual to produce different phenotypes under different conditions. Non-genetic potential variability within the range of a single individual.

Phi scale. Scale used for measuring the grain size of sediments. = -log2 (grain diameter)

Photic zone. The depth zone in the ocean extending from the surface to that depth permitting photosynthesis

Photorespiration. Enhanced respiration of plants in the light relative to dark respiration

Photosynthate. A substance synthesized in the process of photosynthesis

Photosynthetic quotient. In photosynthesis, the moles of oxygen produced, divided by the moles of carbon dioxide assimilated

Photosynthetic rate. The rate of conversion of dissolved carbon dioxide and bicarbonate ion to photosynthetic product

Phototactic. Moving in response to light

Physiological race. A geographically defined population of a species that is physiologically distinct from other populations

Phytodetritus. Particulate organic matter settling through the water column that derives from dead phytoplankton.

Phytoplankton. The photosynthesizing organisms residing in the plankton

Planktivorous. Feeding on planktonic organisms

Plankton. Organisms living suspended in the water column and incapable of moving against water currents

Planktotrophic larva. Planktonic-dispersing larva that derives its nourishment by feeding in the plankton

Planula. The planktonic larval form produced by scleractinian corals and coelenterates

Plate. Major section of the earth's crust, bounded by such features as mid-ocean ridges

Pleistocene. Period of time, going back to approximately 2 million years before the present, in which alternating periods of glaciation and deglaciation have dominated the earth's climate

Pleuston. Refers to plankton that have a float protruding above the sea surface, such as the Portugese man-of-war.

Poikilotherm. An organism whose body temperature is identical to that of the external environment

Point source. A pollution source from a confined spot, such as an industrial pipe.

Polyp. An individual of a solitary coelenterate or one member of a coelenterate colony

Polyphyletic. Refers to a group of species that do not have one common ancestor species

Population density. Number of individuals per unit area or volume

Pop-up satellite archival tags. Fish tags designed to become detached from fish after a time, so that they can rise to the surface and be detected by a global positioning system.

Porifera. The phylum comprising the sponges.

Pound net. A blind-ended net that captures fish at the blind end.

PPT. A measure of the salt content of sea water in terms of kg salt per kg of water, reckoned in parts per thousand. A conductivity version of this measure is the PSU, which differs from PPT by very little, on the order of 0.02 PSU or less.

Precision. Precision is the repeatability of a measurement. A measurement can be precise, but not accurate. See Accuracy and Resolution.

Predation. The consumption of one organism by another

Predator. An organism that consumes another living organism (carnivores and herbivores are both predators by this definition)

Primary producer. An organism capable of using the energy derived from light or a chemical substance in order to manufacture energy-rich organic compounds

Primary production. The production of living matter by photosynthesizing organisms or by chemosynthesizing organisms. Usually expressed as grams of carbon per square meter per year

Province. A geographically defined area with a characteristic set of species or characteristic percentage representation by given species

Protein polymorphism. Presence of several variants of a protein of a given type (e.g., a certain enzyme, such as carboxylase) in a population

Pseudofeces. Material rejected by suspension feeders or deposit feeders as potential food before entering the gut

PSU. Practical Salinity Units. A measure of the salt content of seawater (practical salinity), based upon electrical conductivity of a sample relative to a reference standard of sea water, which now happens to be a reference set of diluted sea water samples from the North Atlantic of known salt content. See PPT.

Pteropods. Group of holoplanktonic gastropods

Purse seine. A seine net that is set in a ring and can be reduced in diameter by pulling a line, trapping fish.

Pycnocline. Depth zone within which sea-water density changes maximally

Q10 Increase of metabolic rate with an increase of 10ºC

Quantitative genetics. The study of the genetic basis of traits, usually explained in terms of the interaction of a group of genes with the environment

Radiocarbon technique (primary productivity). The estimation of primary productivity by the measurement of radiocarbon uptake

Radiolaria. Protistan phylum, whose members are planktonic and secrete an often elaborate siliceous test

Radula. A belt of teeth, found in gastropods and chitons. Used for feeding.

Random spatial distribution. Situation in which individuals are randomly distributed in a space; probability of an individual's being located at any given point is the same irrespective of location in the space

Recruitment. The residue of those larvae that have: (1) dispersed; (2) settled at the adult site; (3) made some final movements toward the adult habitat; (4) metamorphosed successfully, and (5) survived to be detected by the observer

Redox-potential discontinuity. That depth below the sediment-water interface marking the transition from chemically oxidative to reducing processes

Red tide. A dense outburst of phytoplankton (usually dinoflagellates) often coloring water red brown

Refuge. A device by which an individual can avoid predation

Regeneration production. Primary production that is caused by nutrients that are recycled within a water body from excretions.

Regulator. An organism that can maintain constant some aspect of its physiology (e.g., body temperature) constant despite different and changing properties of the external environment

Renewable resource. A resource that can be regenerated (e.g., a growing diatom population that is being exploited by a copepod)

Reproductive effort. The fraction of assimilated nutrients that are devoted to reproductive behavior and gamete production

Residence time. The time a unit of a substance spends in a specified location in the environment (e.g., the residence time of sodium in the Antarctic bottom water).

Resilience. The capacity of a community or ecosystem to respond to a major disturbance and recover quickly to its former state.

Resolution. The smallest amount of change that an instrument can discriminate.

Resource. A commodity that is required by an organism and is potentially in short supply

Respiration. Consumption of oxygen in the process of aerobic metabolism

Respiratory pigment. A molecule, polymer, or other complex adapted to bind and transport oxygen efficiently, usually in a circulatory system (e.g., hemoglobin)

Respiratory quotient. The ratio of moles of carbon dioxide produced to oxygen consumed in respiration

Rete mirabile. A countercurrent exchange structure of capillaries that allows gas uptake in a fish swim bladder

Reverse Bohr effect. Effect that occurs when lactate builds up in the blood of certain invertebrates and pH decreases, increasing the affinity of hemocyanin for oxygen

Reynolds number. A number that represents the relative importance of viscous forces and inertial forces in a fluid. As Re increases, inertial forces become more important. In sea water, Re increases with increasing water velocity and with the size of the object in the water

Rip current. A concentrated rapid current moving offshore from a beach fronting a longshore current

Rise. Bottom of low relief at the base of the continental slope

ROV. Abbreviation for remotely operated vehicle, usually a submersible tethered to a ship, with facilities for video, remote sampling by grabbing arms, and precise navigation

Salinity. Number of kg of dissolved salts in one kg g of seawater, measured in parts per thousand. Actually this definition stands alongside another definition based upon water standards whose electrical conductivities are measured. See PPT and PSU.

Salps. A group of pelagic tunicates (phylum Urochordata), either colonial or solitary, with buccal and atrial siphons on opposite sides of the body

Salt glands. In mangroves: in some species, glands in the leaves that excrete salts to the leaf surface.

Salt marsh. A coastal habitat consisting of salt-resistant plants residing in an organic-rich sediment accreting toward sea level

Saxitoxin. A neurotoxin that blocks sodium channels, produced by certain phytoplankton species.

Scaphopoda. Class of the phylum Mollusca with elongate tusk-shaped conical shell.

Scavenger. An organism that feeds on dead or decomposing animals or macrophytes

Scleractinia. Order of coelenterates, usually producing calcareous skeletons with hexameral symmetry

Scope for activity. The surplus of energy available for activity, such as swimming, beyond that required for maintenance.

Scope for growth. The surplus of energy available for growth beyond that required for maintenance

Scyphozoa. the true jellyfish, members of the phylum Cnidaria

Sea-floor spreading. The horizontal movement of oceanic crust

Seasonal estuary. An estuary in which salinity at any one geographic point changes seasonally (e.g., decreases during the spring melt)

Seaward. Side of an island that faces the direction of wave action generated either by winds or by currents generated by more indirect forces

Secondary production. The production of living material per unit area (or volume) per unit time by herbivores. Usually expressed as grams carbon per meter square per year

Seine net. Net placed in the water and pulled along, capturing marine organisms in the mesh.

Selection. A change in allele frequency over time in a population

Semi-infaunal. Living partially buried within the sediment but partially projecting into the water column.

Sequential hermaphrodite. An individual that sequentially produces male and then female gametes or vice versa

Sessile. Immobile because of an attachment to a substratum

Seston. Particulate matter suspended in seawater

Setules. Chitinous projections from copepod maxillipeds that trap food particles

Shelf-slope break. Line marking a change from the gently inclined continental shelf to the much steeper depth gradient of the continental slope

Sibling species. Closely related species that are so similar that they are nearly indistinguishable morphologically

Side-scan sonar. A sonar system producing sound energy that bounces off the seafloor and is subsequently picked up by a detector. The signal gives a picture of the seabed surface, revealing a variety of surface sedimentary features. See also Multibeam sonar.

Sigma. Parameter expressing the seawater density: and equal to 1 minus the density of seawater, measured at a given temperature and at a pressure of l atmosphere

Siphonophores. A group of specialized hydrozoan cnidarians, consisting of large planktonic polymorphic colonies

Sled. A benthic sampling device designed to slide along the sediment surface, digging into the bottom to a depth of at most a few centimeters

Slope. A steep-sloping bottom extending seaward from the edge of the continental shelf and downward toward the rise

Solubility product constant. At saturation, a constant, which is a multiple of the solubilities of the two ionic components (e.g., Na^+ and Cl^-) of the substance that could crystallize in solid phase.

Somatic growth. Growth of the body, exclusive of gametes

Sorting (of a sediment). The range of scatter of particle sizes about the median grain size of a sediment

Space limited. Description of a situation in which space is a limiting resource

Spatial autocorrelation. A situation in which some parameter at any location (e.g., population density) can be predicted through a knowledge of the values of the parameter in other locations

Spatial distribution. The arrangement of individuals in a space

Speciation. The process of formation of new species

Species. A population or group of populations that are in reproductive contact but are reproductively isolated from all other populations

Species-area effect. A regular logarithmic relationship between the number of species in a confined geographic area (e.g., an island) and the area in which the species occur

Species richness. The number of species in an area or biological collection

Sporophyte. Diploid stage in the life cycle of a plant

Spring diatom increase. The major rapid population increase of diatoms, occurring in the spring in temperate-boreal latitudes

Spring tides. Fortnightly tides occurring when the vertical tidal range is maximal

Stability-time hypothesis. Hypothesis that states that higher diversity occurs in habitats that are ancient and stable environmentally

Standing crop. The amount of living material per unit area or volume; may be expressed as grams of carbon, total dry weight, and so on

Stock recruitment models. Fishery models that predict the amount of juvenile recruitment as a function of the parent stock

Stratification. In benthos, the presence of different infaunal species at distinct respective horizons below the sediment-water interface

Subtropical. Refers to the portion of the temperate zone closest to the equator

Succession. A predictable ordering of a dominance of a species or groups of species following the opening of an environment to biological colonization

Surface layer. The layer of the ocean extending from the surface to a depth above which the ocean is homogeneous due to wind mixing

Survivorship curve. The curve describing changes of mortality rate as a function of age

Suspension feeder. An organism that feeds by capturing particles suspended in the water column

Swash rider. Invertebrate that can migrate up and down shore with the rising and falling tide, in order to maintain station at a level that is moist but not overly washed by the waves

Teleplanic larva. Larva capable of dispersal over long distances, such as across oceans

Temperate. Pertaining to the latitudinal belt between 23º 27' and 66º 33' north or south latitude

Tentacle-tube-foot suspension feeder. Suspension feeder that traps particles on distinct tentacles or tube feet (in echinoderms)

Terminal anchor. In hydraulically burrowing organisms: any device used to anchor the leading portion of the burrower, permitting muscular contraction to drag the rest of the body into the sediment

Territoriality. Defense of a specified location against intruders

Tertiary production. The production of living material per unit area (or volume) per unit time by organisms consuming the herbivores. Usually expressed as grams carbon per meter square per year

Thermocline. Depth zone within which temperature changes maximally

Thermohaline circulation. Movement of seawater that is controlled by density differences that are largely explained in terms of temperature and salinity

Tidal current. A water current generated by regularly varying tidal forces

Tides. Periodic movement of water resulting from gravitational attraction between the earth, sun, and moon

Top-down control. Refers to food webs where control of a population is mainly explained by consumption by a species or group of species at higher levels of the food chain (e.g., population change of population of mussels controled by seastar predation).

Trade winds. Persistent winds at low latitudes in both the Northern and Southern hemispheres, blowing toward the west and the equator

Trench. Deep and sinuous depression in the ocean floor, usually seaward of a continental margin or an arcuate group of volcanic islands

Trophic level. In a food chain, a level containing organisms of identical feeding habits with respect to the chain (e.g., herbivores)

Tropical. Being within the latitudinal zone bounded by the two tropics (23º 27' north and south latitude)

Turbidity. The weight of particulate matter per unit volume of sea water

Ultraplankton. Planktonic organisms that are less than 2 micrometers in size

Uniform spatial distribution. Situation in which individuals are more evenly spread in space than would be expected on the basis of chance alone

Upwelling. The movement of nutrient-rich water from a specified depth to the surface

Vertically homogeneous estuary. An estuary in which, at any given location, wind or tidal mixing homogenizes salinity throughout the water column

Vitamin. Chemical substances required in trace concentrations acting as a cofactor with enzymes in catalyzing biochemical reactions

Viviparous (development). Refers to development of an organism through the juvenile stage within a parent

Wash zone. The depth zone in which sediments are disturbed by wave action near the shoreline

Water mass. A body of water that maintains its identity and can be characterized by such parameters as temperature and salinity

Watershed. The land area that is drained by a river or estuary and its tributaries

Westerlies (prevailing westerlies). Persistent eastward-equatorward winds in midlatitudes in both the Northern and Southern Hemispheres

Windward [side]. The side of an island that faces a persistent wind

Within-habitat comparison. A contrast of diversity between two localities of similar habitat type

Wrack zone. A bank of accumulated litter at the strandline

Year-class effect. The common domination of a species population by individuals recruited in one reproductive season

Zonation. Occurrence of single species or groups of species in recognizable bands that might delineate a range of water depth or a range of height in the intertidal zone

Zooplankton. Animal members of the plankton

Zooxanthellae. A group of dinoflagellates living endosymbiotically in association with one of a variety of invertebrate groups (e.g., corals).

INDEX

www.ingramcontent.com/pod-product-compliance
Lightning Source LLC
LaVergne TN
LVHW062222200926
848484LV00006B/1454